A NATURALIST'S GUIDE TO THE
BIRDS OF
MALAYSIA

Geoffrey Davison & Yeap Chin Aik

JOHN BEAUFOY PUBLISHING

Reprinted in 2019
This edition first published in the United Kingdom in 2018 by John Beaufoy Publishing
11 Blenheim Court, 316 Woodstock Road, Oxford OX2 7NS, England
www.johnbeaufoy.com

10 9 8 7 6 5 4 3 2

Photo Credits
Front cover: *main image* Bornean Bristlehead (Cede Prudente), *bottom left* Blue-headed Pitta (Cede Prudente, *bottom centre* Fluffy-backed Tit-babbler (David Bakewell), *bottom right* Silver-eared Mesia (Ooi Beng Yean).
Back cover: Oriental Pied Hornbill (David Bakewell). **Title page:** Little Spiderhunter (Cede Prudente).
Contents page: Black-and-yellow Broadbill (Cede Prudente).
Main descriptions: photos are denoted by a page number followed by t (top), b (bottom), l (left) or r (right).
David Bakewell: 12b, 15t, 16tr, 18t, 18b, 20t, 20m, 20b, 21t, 21b, 22t, 22b, 26tl, 27t, 28t, 28b, 29b, 31t, 32b, 34t, 35t, 35b, 36tl, 36tr, 36b, 37t, 37b, 38t, 38b, 39t, 39b, 40t, 41t, 41b, 42b, 43b, 47t, 48bl, 49t, 53t, 53b, 55t, 56t, 57t, 57b, 58t, 58b, 59t, 62t, 63t, 64b, 65t, 68t, 71t, 72t, 78b, 81t, 81b, 83t, 83b, 85t, 87t, 87b, 89t, 89br, 90b, 93t, 93b, 94b, 96t, 97t, 97b, 100tl, 100tr, 100b, 103t, 103b, 104t, 104bl, 104br, 105tr, 106br, 107b, 109b, 110b, 111b, 113t, 113b, 116t, 116b, 117t, 117b, 118t, 120tl, 120tr, 121bl, 122t, 123t, 124t, 128b, 131t, 132t, 133t, 133b, 134t, 134b, 135t, 135b, 136t, 136b, 137t, 137b, 138t, 138b, 139t, 141t, 141b, 142t, 142b, 143t, 144t, 144b, 145t, 145b, 146t, 147t, 147b, 148t. **John Corder:** 14t. **Rosmadi Hassan:** 151t. **David Lai:** 13m, 14b, 17b, 32t, 33t, 33b, 40b, 42t, 43t, 44b, 45t, 45b, 50t, 67b, 70b, 74b, 84t, 91t, 92tl, 99t, 107tl, 126b, 127t, 127b, 149t. **J. Cede Prudente:** 13b, 16b, 17t, 24t, 30b, 44t, 48t, 55b, 62b, 66t, 73t, 82t, 82b, 92tr, 95t, 95b, 99b, 102b, 106tr, 108b, 112t, 114b, 115t, 119b, 139b, 146b. **Ooi Beng Yean:** 12t, 13t, 15b, 16tl, 19t, 19b, 23t, 23b, 24b, 25t, 25b, 26tr, 26b, 27b, 29t, 30t, 31b, 34b, 46t, 46b, 47b, 48br, 49b, 50b, 51t, 51b, 52t, 52b, 54t, 54b, 56b, 59bl, 59br, 60t, 60b, 61t, 61b, 63b, 64t, 65b, 66b, 67t, 68b, 69t, 69b, 70t, 71b, 72b, 73b, 74t, 75t, 75b, 76t, 76b, 77t, 77b, 78t, 79t, 79b, 80t, 80b, 84b, 85b, 86tl, 86tr, 86b, 88t, 88b, 89bl, 90t, 91b, 92b, 94t, 96b, 98t, 98b, 101t, 101b, 102t, 105tl, 105b, 106tl, 106bl, 107tr, 108t, 109t, 110t, 111t, 112b, 114t, 115bl, 115br, 118b, 119t, 120b, 121t, 121br, 122b, 123b, 124b, 125t, 125b, 126t, 128t, 129t, 129b, 130t, 130b, 131b, 132b, 140t, 140b, 143b, 148t, 149b, 150t, 150b, 151b.

-912081-63-9

esigned and typeset by D & N Publishing, Baydon, Wiltshire, UK

d and bound in Malaysia by Times Offset (M) Sdn. Bhd.

·CONTENTS·

ACKNOWLEDGEMENTS

Thanks go to Ken Scriven, the generosity of the photographers, David Bakewell, David Lai, Ooi Beng Yean, Cede Prudente and John Corder, and to colleagues at the National Parks Board, Singapore, and the Malaysian Nature Society, including Jessie Ong.

INTRODUCTION

Five important land masses – the Malay Peninsula, Sumatra, Borneo, Palawan, and Java and Bali combined – make up the biogeographical area known as Sundaland, in Southeast Asia. This area is characterised by its humid equatorial climate, maritime climatic influence, limited seasonality and climax vegetation of tropical evergreen rainforest. Five countries are covered by Sundaland: Malaysia, Singapore, Indonesia, Brunei and the Philippines. This book introduces the birds of Malaysia and Singapore, many of which are shared with the other three countries in this diverse and fascinating area. Peninsular Malaysia is the southernmost part of mainland Asia, while the two big Malaysian states of Sabah and Sarawak lie on the island of Borneo.

CLIMATE

Sundaland lies within the Intertropical Convergence Zone, where the prevailing directions of the northeast monsoon (November–April) and southwest monsoon (May–October) are modified by the region's proximity to the Equator. Strictly, this is not a monsoon zone but an inter-monsoon zone. Peak rainfall tends to be at the changeover from one monsoon direction to the other, and may occur once or twice a year depending on site. Annual rainfall totals vary from around 1,500mm in drier localities within the lowlands, to over 3,000mm in wetter localities and some montane areas. The wettest weather typically occurs between October and January, when there are more gloomy days and consistent rain; if there is a secondary rainfall peak around April–May, it tends to consist of fewer, shorter, more intense storms. Late-afternoon rainfall is typical year-round, but heavy storms can roll in from the sea in the early morning. Although this is the general picture, many local variations occur, and as elsewhere in the world the weather can do virtually anything at any time.

Daily maximum temperatures in the lowlands are around 32–35°C; night minima are around 24°C in urban areas, but inside undisturbed lowland forest they can be as low as 17°C. Mountains are correspondingly cooler, with a fall in night-time temperatures of about 1°C per 250m altitude. Hail is a rarity, provoking newspaper comment, and short-lived, thin ice has been recorded near the summit of Mount Kinabalu. The drought in 1982–83 changed perceptions of weather patterns for ever: several major El Niño years have occurred since then, with consequent fires and 'haze'.

El Niño

In weather terms, no two years are alike, and weather patterns and seasons are based on statistical averages. There are daily, seasonal and superannual statistical changes, of which the El Niño Southern Oscillation (ENSO) is one of the most notorious. This phenomenon is caused by atmospheric pressure differences between the eastern and western Pacific, resulting from alterations every few years in sea currents and hence sea temperature, air temperature and rainfall. The changes in the eastern Pacific during severe El Niño years (cooler temperatures and usually wetter weather) are quite different from those in the western Pacific, including Southeast Asia, where several months of drought can occur. An El Niño 'year' can span parts of more than one calendar year, and is typically followed by a wetter-than-usual La Niña year in the western Pacific. The effects of the ENSO on the forest, the birds and other wildlife of Sundaland are profound.

VEGETATION

The climatic climax vegetation throughout the region is tropical evergreen rainforest. Semi-evergreen (or semi-deciduous) forest occurs only further north in Thailand, and in drier, more seasonal parts of Java and lands to the east. Rainforest is a general term, covering various forest types growing in areas where rainfall typically exceeds evapotranspiration for at least nine months of the year. (There are many other definitions in the literature.) The classical progression from the coast to inland is from mangroves to peat-swamp forest, freshwater swamp forest, dryland lowland dipterocarp forest, hill dipterocarp forest, lower montane forest and upper montane forest. Above this, only one mountain in Malaysia, Mount Kinabalu (4,095m), reaches beyond the tree-line, although there are others nearby in Sumatra and Java.

Only about 30–35 plant species constitute 'true' mangroves (plants not able to survive in other forest types). Low species diversity, fairly simple vegetation structure with a uniform canopy, lack of understorey and daily inundation by tides create a demanding environment. Woodpeckers such as the Sunda Pygmy (p. 75) and Common Flameback (p. 77) cope well here. The Mangrove Pitta, like the Blue-winged (p. 81) but with a heftier bill, is a ground feeder in mangroves, and must retreat with the tide and find a secure mound above water level for its leaf-bundle nest. The dried mounds created by mud-lobsters – home also to a species of trapdoor spider and other exotic creatures – provide possible sites. Herons and egrets, foraging on adjacent mudflats at low tide, retreat to mangroves at high tide and at night: Little Herons (p. 20) and Little Egrets (p. 22) tend to roost on the stilt roots and lower branches, while Great Egrets (p. 22) and Grey and Purple herons (p. 23) favour the tree crowns. The region's two species of night-heron, including the Black-crowned (p. 19), are both specialist mangrove nesters.

Lesser Adjutants (p. 17) and many shoreline waders (pp. 34–40) feed at pools in clearings in mangroves. Sites such as Bako-Buntal Bay (Sarawak) and the Kuala Selangor and Matang mangroves (Peninsular Malaysia) are famed for the large numbers of waders that can be seen on their huge exposed mudflats. Various wader species forage on sandy or muddy areas of differing wetness and consistency. They specialise in different species of burrowing worms, small molluscs or crustaceans, excavated at different depths below the surface according to the species' preference and bill length, which determines how deeply the birds can probe for food.

Lowland rainforest (from the extreme lowlands up to about 900m altitude) is the premier birding habitat, with high species diversity and classic Southeast Asian groups such as leafbirds, barbets and hornbills. Undisturbed forest has a complex three-dimensional structure, the layers often recognised as understorey, lower storey, middle storey and canopy, with scattered emergent trees overtopping the rest at heights of up to 60–70m. The forest is usually dominated by large timber trees of the family Dipterocarpaceae (dipterocarps), but these provide little food for wildlife. Important fruiting trees are laurels, mangoes and mangosteens, among many others, and for birds especially the vast array of figs (more than 100 species), of which at least some species are in fruit at any one time. Statistically identifiable fruiting seasons, and the great distances between individual trees of any given species, mean that large frugivores (fruit-eaters) such as Great and Rhinoceros hornbills (p. 70) must often travel long distances over the forest canopy. By contrast, small understorey insectivores (insect-eaters) such as the Scaly-crowned Babbler (p. 143) and Short-tailed Babbler (p. 144) have small home ranges. The forest is very heterogeneous, with, for example, White-chested Babblers (p. 144) near riverbanks, Velvet-fronted Nuthatches (p. 112) on the main trunk and larger boughs of trees in the upper storey, Abbott's Babblers (p. 143) in swampy forest and forest edges, and Little Spiderhunters (p. 101) in old treefall areas with banana regrowth. Many more notes are needed from naturalists on the detailed habitat preferences of each species.

Zones of altitudinal change from lowlands to lower (roughly 900–1500m) and upper montane forest (roughly 1,500m upwards) are of great interest. The character of these zones varies from place to place depending on topography and rainfall, tree species composition and also the species of birds present. The Velvet-fronted Nuthatch (p. 112), for example, ranges higher in Sabah and Sarawak (where the montane Blue Nuthatch, p. 112, is absent) than it does in Peninsular Malaysia.

BIOGEOGRAPHY

Repeatedly in the following pages are described species that are found 'from the Himalayas though southern China and Southeast Asia' to Peninsular Malaysia, Sumatra, Borneo and Java. This area represents the eastern half of the Oriental region, otherwise known as the Indo-Malayan region, whereas its western half is made up primarily of Pakistan, India, Sri Lanka and Bangladesh. The Indo-Malayan region is one of the world's great geographical zones and is crucial to biological diversity. Largely covered in tropical rainforest, its total flora and fauna are vast, and include some of the world's most charismatic large land mammal species, such as the Asian Elephant *Elephas maximus*, the Bornean and Sumatran orang-utans *Pongo pygmaeus* and *P. abelii*, and the Javan Rhinoceros *Rhinoceros sondaicus* and Sumatran Rhinoceros *Dicerorhinus sumatrensis*. Also included are pheasants (pp. 13–14), raptors (pp. 24–30), doves and pigeons (pp. 43–46), hornbills (pp. 69–71) and barbets (pp. 72–74) in great array. Many species are confined to Peninsular Malaysia, Sumatra and Borneo, the ever-humid core of Sundaland, and are particularly vulnerable to forest loss. The forest here contains more than 300 species of dipterocarp trees, many of them endemic, and the total flora comprises well over 15,000 plant species for the whole region.

Some birds of mainly Himalayan or at least continental Asian distribution reach no further than Peninsular Malaysia, such as the Blue-winged Siva (p. 147). Others may reach as far as Sumatra, such as the Silver-eared Mesia and Long-tailed Sibia (p. 148). Although these species can be regarded as lowland or foothill birds in the northern part of their range, they are confined to montane forest further south.

Forest specialisation is especially true of the species found in the more restricted region of Sundaland. Here, species are characteristically confined to the undisturbed forest. Thus, for example, the Coppersmith Barbet (p. 74), found from the Himalayas through southern China and Southeast Asia to the Philippines, but missing from forest-rich Borneo, is a bird of semi-open country and park-like woodlands with spacing between the main trees. In contrast, the Yellow-crowned Barbet (p. 74) is confined to the Sunda Shelf, and is exclusively found in the canopy of deep forest.

Most of the endemic birds of Sundaland are forest birds, and most are confined to mountains. Examples are the Chestnut-crested Yuhina (p. 139) of Borneo, and the Sunda Laughingthrush (p. 146) of Borneo and Sumatra. The Dusky Munia (p. 108) is one of the few open-country lowland Sunda endemics, in this case confined to Borneo.

Where birds are found extensively both in mainland Asia and Sundaland, some large genetic differences have been detected between populations north and south of Peninsular Thailand. It is risky to be too precise about the transition point, which is often pinpointed as the Isthmus of Kra; reality is often more complex and fuzzy. The Little Spiderhunter (p. 101) is such a bird, showing large genetic differences but almost no plumage distinctions between northern and southern Asian populations.

The notes on distribution in this book are generalisations. They are not meant to define the past or present distribution of the species very precisely, so they often miss out details of the smaller islands, and in particular omit places where the birds once occurred but no longer do so. All records of birds on islands are of potential scientific value, and need to be recorded meticulously. On the mainland, birdwatchers tend to visit just a handful of well-known, accessible places repeatedly, so new locality records from a wider spread of visits will be very valuable.

Climate, vegetation and biogeographical conditions conspire every two to seven years to produce mass fruiting events, during which most dipterocarps and many other trees in the lowland forest fruit, or fruit more heavily. This provides much more food than usual for many animals, and bird activity and breeding tend to be more obvious at such times. Mass fruiting may be triggered by a succession of cool nights during dry weather, and may also be linked to El Niño climatic events, although this is still unclear. The droughts associated with powerful El Niños can be times of food shortage and stress among many birds, displacing them to unusual localities, altitudes and habitats, and sometimes forcing them to take food items they would not normally touch.

NESTS

Many forest birds make tiny nests, designed to be inconspicuous to predators, such as those of the White-throated Fantail (p. 91) and Black-winged Flycatcher-shrike (p. 84). Nests may be incredibly well camouflaged: the nest of Gould's Frogmouth (p. 56), for example, is particularly hard to spot and the plumage of the incubating bird makes it harder still. Hanging the nest in an inaccessible position is typical of the Black-naped Monarch and Blyth's Paradise-Flycatcher (p. 93), and also of the Black-and-red Broadbill (p. 80), whose nest is hung over water and looks like a tangle of litter left behind by a flood.

In forest, the abundance of trees makes cavity nesting an obvious strategy. Barbets (pp. 72–74) and woodpeckers (pp. 75–77) make their own cavities in rotten or slightly rotting wood, whereas Blue-eared Kingfisher (p. 66) makes a burrow in a forested riverbank, and Collared Kingfisher (p. 65) may nest in a mangrove bank, roadside cutting or hollow tree cavity, and perhaps even in an arboreal termites' nest. Trogons (pp. 60–62) use natural cavities in rotten stumps in the forest, whereas hornbills (pp. 69–71) need to seek out cavities that are roomy, safe and high enough inside large living trees. It is sometimes possible to observe nests of large raptors, such as White-bellied Fish-eagle (p. 28), or Blyth's Hawk-eagle (p. 29) from a distance, and this gives plenty of scope for naturalists to study nest-building and incubation, care of the young, and the amount and type of food brought to the nest.

In spite of much birdwatching in the region, there are still dozens of species whose nests are known from only a handful of examples, and others whose nests have never been found. All observations of nests, including the date, position, construction materials and other details, should be recorded.

Tides, Seashores and Waders

Most of the 60 or so waders that occur in Southeast Asia are widespread, and are well known through much of Europe and Asia: Whimbrel (p. 37), Common Greenshank (p. 38) and Common Redshank (p. 39) are familiar examples. A few, such as Nordmann's Greenshank *Tringa guttifer* and Spoon-billed Sandpiper *Eurynorhynchus pygmeus* (not included here), are special and rare, with very restricted breeding grounds and small numbers occurring unpredictably at a few wintering sites in Peninsular Malaysia and Singapore.

The west coast of Peninsular Malaysia (around Matang, Pulau Kelang and Kuala Selangor), and some spots in western Sarawak (Bako-Buntal Bay and Pulau Bruit), have extensive mudflats that attract huge numbers of passage and winter migrants. Low tides occur once or twice a day, depending on locality, often leaving the mud exposed mainly at night and in the early morning. Ringing has shown the importance of a continuous chain of feeding sites at which migrants can refuel en route from the Siberian tundra and Korea southwards, including to Australasia. Coastal reclamation, bunds, industrial and residential development, and pollution are all potential pressures on most such sites.

Wader concentrations are less well known in Sabah, where the extensive mudflats of the east coast (Labuk, Darvel Bay and Tawau) are less accessible to birdwatchers. Investigation of these areas could reveal important bird sites, not only for waders but also for herons and egrets. Only one wader breeds locally, the Malaysian Plover *Charadrius peronii*, which is confined to sandy beaches and hence is being impacted by seafront development and tourism.

Seabirds

There's no doubt that seabirds in Southeast Asia are severely underwatched. Not many birdwatchers go out to sea for long periods, and while scuba-diving is increasingly popular most divers are not birdwatchers. There are few seabird nesting islands in the region. Pulau Perak (between the Malay Peninsula and Sumatra) and Pulau Layang-Layang north of Sabah are the best known, but only Layang-Layang is easily accessible. There are other very important seabird nesting islands in the Spratly Group, at Tubbataha Reef in the Philippines and elsewhere, but these are also remote. Birdwatchers tend to see single birds or small flocks foraging far from their nesting and roosting sites, and notes about these sightings are worryingly scarce. The distribution and seasonality of seabirds in the region (mostly terns, but also gulls, skuas and jaegers, petrels and shearwaters) are all poorly understood. Sea and weather conditions, foraging habits and movements all require investigation. New records of many species are possible.

Opportunities for Naturalists

Local and foreign birdwatchers tend to visit a few well-known sites in the region. However, there is a lot of scope for visiting new places and expanding the bare list of localities from which many birds have been recorded. Upper and lower altitude limits have become a significant concern in detecting the impacts of climate change, but much caution needs to be exercised in recording altitudes (roadsigns can be highly misleading and maps unreliable). Due recognition also needs to be given to the point that the sport of finding the highest or lowest individual tends to emphasise the exceptional bird, not the population norm. Not too much should be read into single records.

The spread of recent invaders is sometimes poorly recorded, because these can be common birds of little interest to birdwatchers, such as the Javan Myna (p. 113) or Eurasian Tree-sparrow (p. 110). The Collared Kingfisher (p. 65) and Oriental Magpie-robin (p. 123) have both expanded their ranges inland from mangrove coastal habitat into woodland, agricultural estates and open country, and similar changes leading to spread through new habitats could be expected in other species.

Records from islands can give clues about colonisation ability, and if repeated they may show species turnover rates. Records from many mountains, still with poor coverage, may indicate distance from other known populations, faunal composition, and ability to survive in small habitat patches.

Group size, height in the vegetation at which a species occurs and normal foraging behaviour are not well described for many birds. How the different tailorbirds (pp. 149–150), flycatchers (pp. 118–122) or forest babblers (pp. 140–145) partition their habitat could be very interesting. Sometimes, species-specific habitat differences (such as between mangrove, riverine and inland habitats) seem clear, but in places the distinctions appear to break down; there could be many subtleties in avoidance of competition.

Migration dates, numbers and directions should be recorded. Very few observations have been published on migrant landfall on the Sarawak and Sabah coasts, where the arrival of waders, flycatchers, warblers and raptors should be of great interest.

In short, there is a lot for naturalists to do besides seeking out rarities or compiling checklists, and almost every day out should result in small facts that will add to the overall knowledge of the bird species found in this fascinating area.

WHERE TO GO

Rather than describing a few places, with their attendant birds and details of how to get there, a list of place-names is provided below that can be researched on websites or in previous site guides. For more detailed descriptions and lists, but rather outdated information, read Bransbury (1993); and for online site reports, visit www.travellingbirder. com or www.eurobirding.com.

Peninsular Malaysia
Matang; Taman Negara National Park; Krau Wildlife Reserve; Kuala Selangor Nature Park; Tanjung Karang coast and rice fields; Kuala Gula; Tasik Bera and Tasik Chini; Fraser's Hill; Cameron Highlands; Genting Highlands, Gunung Bunga Buah and Old Gombak Road; Pekan and Nenasi peat-swamp forest; Belum-Temengor forest; Panti forest.

Singapore
Sungei Buloh Wetland Reserve; Pulau Ubin; Bukit Timah Nature Reserve; Central Catchment Nature Reserve; Pulau Semakau; Changi Village and Changi coast; Kranji Reservoir and Neo Tiew.

Sarawak
Bako-Buntal Bay; Semengoh; Borneo Highlands Resort; Bako National Park; Gunung Mulu National Park; Bario and the Kelabit Highlands.

Sabah
Tunku Abdul Rahman Park; Long Pasia (Ulu Padas); Kampung Benoni; Likas Lagoon (Kota Kinabalu Wetlands Centre); Tempasuk Plain and lagoon; Kinabalu Park; Sukau and Kinabatangan River; Danum Valley; Tabin Wildlife Reserve; Semporna and its offshore islands; Pulau Layang-Layang; Sepilok forest.

WHERE TO SUBMIT RECORDS

Bird records are an important source of information in helping to understand more fully the avifauna of this region. For Malaysia, birdwatchers are encouraged to submit their sightings to the MNS-Bird Conservation Council electronically via the bird-i-witness database at www.worldbirds.org/malaysia or email mnsrc.rc@gmail.com. For Singapore, record submissions can be made to the Nature Society (Singapore) Bird Group website at http://wildbirdsingapore.nss.org.sg.

Where to Publish Information

Suara Enggang is the quarterly birding magazine of the Malaysian Nature Society, which accepts short articles on sightings and birdwatching. The society also publishes the quarterly magazine *Malaysian Naturalist* and the scientific publication *Malayan Nature Journal*. The society has branches in most Malaysian states, including Sabah and Sarawak.

In Singapore, the Bird Group of the Nature Society (Singapore) solicits and reviews sightings, and the Bird Ecology Study Group has an online blog. The society publishes the quarterly *Nature Watch*.

The Oriental Bird Club, based in Sandy, Bedfordshire, publishes *Birding Asia* twice yearly, with summaries of records from many Asian countries, including Malaysia and Singapore; it also published the annual scientific journal *Forktail*.

For contact details of these organisations, see p. 172.

Bird Topography

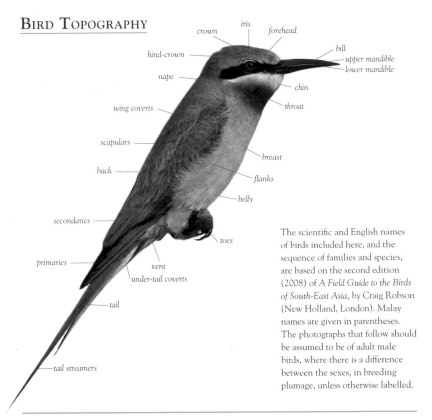

crown · iris · forehead · hind-crown · bill · upper mandible · lower mandible · nape · chin · wing coverts · throat · scapulars · breast · back · flanks · belly · secondaries · toes · primaries · vent · under-tail coverts · tail · tail streamers

The scientific and English names of birds included here, and the sequence of families and species, are based on the second edition (2008) of *A Field Guide to the Birds of South-East Asia*, by Craig Robson (New Holland, London). Malay names are given in parentheses. The photographs that follow should be assumed to be of adult male birds, where there is a difference between the sexes, in breeding plumage, unless otherwise labelled.

Barred Buttonquail ▪ *Turnix suscitator* (Burung Puyuh Tanah) 16cm

DESCRIPTION Placed here for ease of comparison, but not related to quails and partridges. Stocky ground-bird whose breast and flanks are strongly marked with black lozenge-shaped bars, upperparts scaled and mottled, belly and under-tail coverts a rich

cinnamon. Female (shown here) has black throat and upper breast, absent in duller male. Strong bill and staring yellow eye. **DISTRIBUTION** India through S China to Taiwan and Japan, southwards through Southeast Asia to the Philippines, Sulawesi and Lesser Sundas. Resident in Peninsular Malaysia and Singapore, but absent from Borneo, including Sabah and Sarawak. **HABITS AND HABITAT** Increasingly scarce, but always hard to see even where present, in old agricultural land, thick grass and roadside scrub. Reversed sexual dimorphism, the dull male incubating eggs and tending chicks. Female's call is a series of soft booms, increasing in volume towards the end.

Bornean Partridge ▪ *Arborophila hyperythra* (Kangkerang) 27cm

DESCRIPTION Chubby, secretive brown ground-bird whose flanks are strongly marked with white discs on a black surround; brow and face bright chestnut (with much grey on eyebrow in some populations), crown and eye-line dark brown. Grey bill, red legs and red skin around eye. Sexes are alike. **DISTRIBUTION** Endemic to Borneo, in mountains

at 650–3,050m (usually 900–2,200m). Resident from Mount Kinabalu in Sabah to Mount Mulu and the Kelabit Highlands in Sarawak, and southwards into Kalimantan. **HABITS AND HABITAT** Often heard once the call is known, but seldom seen; appears to be thin on the ground, in montane forest where vegetation is undisturbed, and in dense old regrowth several years after shifting cultivation has ceased. Long, straight claws are used for raking in the leaf litter for food. Pairs duet, 1 bird giving a long series of single piping whistles *too, too, too*, while the other gives 2-note whistles, *kee-too, kee-too*....

Red Junglefowl ■ *Gallus gallus* (Ayam Hutan) 45–75cm

DESCRIPTION Unmistakably chicken-like, male with bright golden-brown neck hackles, black underparts and glossy greenish sickle-shaped tail; female oatmeal-brown or darker, with paler face and darker tail. Breeding males have bright red comb and face, and white earlobes, but in dull eclipse the neck hackles and tail sickles are shed. **DISTRIBUTION** Resident from India to Nepal and S China, through most of Southeast Asia; natural E limit is Java, but now occurs through to the Philippines and Lesser Sundas owing to ancient human intervention. Wild or semi-wild birds are found through much of Peninsular Malaysia and Singapore, and feral birds are now spreading in Sabah. **HABITS AND HABITAT** Typically, single males, or 1 male with a few females, or mated pairs, sometimes with young, forage among low vegetation beneath the trees. Old, poorly maintained rubber estates and oil-palm estates are the favoured habitat, with nests made at tree bases or among dense undergrowth. Male's crow is like that of a domestic cockerel but with a sharp cut-off.

Crested Fireback
■ *Lophura rufa* and *L. ignita* (Ayam Pegar) 65–70cm

DESCRIPTION Very attractive pheasants with diagnostic bushy crest, blue facial skin and pale feet. Adult males from Peninsula have iridescent blue-black plumage with white flank-streaks, maroon and orange rump, and black-and-white vaulted tail; males from Borneo have buff and black tail and chestnut-red belly. Females are mainly brown with white streaks on chest and scales on underparts. **DISTRIBUTION** One species, *L. ignita*, resident in southernmost Myanmar, Malay Peninsula, and Sumatra. The other, *L. rufa*, in Borneo, including both Sabah and Sarawak. Not in Singapore. **HABITS AND HABITAT** Restricted to primary lowland forests, and just into montane forest to 1,200m in places. Like other pheasants, is quite gregarious and travels in small parties in search of fallen fruits, ants and termites. Voice is a range of whirrs and gurgles, and a loud cluck, but no powerful advertising call; males stand and whirr their wings briefly to produce a throbbing sound as advertisement.

ABOVE RIGHT: *Malayan Crested Fireback*. RIGHT: *Bornean Crested Fireback*

Malaysian Peacock-pheasant

■ *Polyplectron malacense (Kuang Bongsu)* Male 50–55cm; female 40–45cm

DESCRIPTION A handsome medium-sized pheasant. Plumage in both sexes is generally dark cinnamon-brown, intricately spotted and vermiculated, the male with many

iridescent green ocelli on the mantle, wings and tail. Bare facial skin bright orange in male, dull orange to grey in female. Adult males sport a short brush-like crest hanging forward over the bill. **DISTRIBUTION** Resident in the Malay Peninsula south of about 12° N. Now possibly extinct in Thailand. Not found in Singapore, or in Borneo where a different species occurs. Resident at low elevations up to 300m. HABITS AND HABITAT Resident of lowland forests, often heard but rarely seen, usually alone or in pairs. Males can be very vocal during the breeding season with haunting two-note whistle, incessant harsh clucking, and angry chicken-like squawks. Males maintain short-term dancing grounds for their displays.

Great Argus ■ *Argusianus argus* (Kuang Raya) Male 160–200cm; female 70–75cm

DESCRIPTION Largest pheasant in the forest, with a bare blue head, red feet and generally brownish plumage in both sexes. Adult males have extraordinarily long central tail feathers and greatly expanded wing feathers with ocelli that are usually hidden from view. Females have shorter tail and

wings, and no ocelli. **DISTRIBUTION** Resident in S Myanmar, Malay Peninsula, Sumatra and Borneo, including Sabah and Sarawak. Not in Singapore, though conceivably once present. **HABITS AND HABITAT** Encounters are confined to pairs or singles in lowland and hilly primary and old secondary forests. Males are territorial and construct dancing grounds that are kept free from leaves, sticks and stones. These grounds are each used by 1 male to stage spectacular displays to attract females. Highly vocal, giving an explosive, repeated kuau that echoes through the forests.

Lesser Whistling-duck
■ *Dendrocygna javanica* (Itik Belibis) 40cm

DESCRIPTION Lightly built duck with a slim neck, the plumage overall buffy brown with brighter chestnut on wing coverts and rump. In flight, wings are black beneath and black with chestnut coverts above. Similar to **Wandering Whistling-duck** *D. arcuata*, but lacks white on rump and long, pale flank plumes of that species. **DISTRIBUTION** Resident in India, Indochina and S China, and throughout Southeast Asia to the Lesser Sundas. Formerly common, but now may be hunted, displaced by removal of habitat and disturbed by human activity. **HABITS AND HABITAT** Usually found in small flocks, although there are old reports of much greater numbers. Within the flock, mated pairs stay together, sometimes performing display rushes across the water, each bird with 1 wing raised. Dives are brief. A tree nester, choosing sites in cavities or clefts between large boughs; ducklings jump to ground after hatching.

Cotton Pygmy-goose ■ *Nettapus coromandelianus* (Itik Kapas) 36cm

DESCRIPTION Small silky white duck, male with green-glossed black cap, breast-band and wings, and broad white wing bar visible in flight. Female is similar but with smoky-grey underparts, less glossy above, and lacking breast-band. **DISTRIBUTION** India and Sri Lanka through S China to Taiwan, southwards through Southeast Asia and the Lesser Sundas to New Guinea and N Australia. Perhaps resident in parts of Peninsular Malaysia, Sabah and Sarawak, but seldom seen and apparently mobile over short or long distances; appears irregularly and patchily in Singapore. **HABITS AND HABITAT** Freshwater ponds and marshes, old mining pools, rarely in brackish ponds; probably more characteristic of the seasonal tropics N and S of region, and not well adapted to ever-humid forested habitats. Large flocks can be seen occasionally in northernmost Peninsular Malaysia; gangs of males pursue females in display flights, giving high-pitched gaggling calls.

LEFT: *Male*. RIGHT: *Female*

Little Grebe ■ *Tachybaptus ruficollis* (Burung Gerib Kecil) 27cm

DESCRIPTION Small, dark waterbird with short bill and dark cap. Plumage generally brown with slight greyish tone, darker on upperparts. Neck and sides of head in breeding birds are bright reddish brown but turn buffy when non-breeding. A prominent light yellowish spot at base of gape and pale iris are useful identification features. **DISTRIBUTION** Widely distributed throughout Africa, Europe, temperate Asia, India, China, and Southeast Asia to New Guinea. Resident and migrant in Peninsular Malaysia and Singapore; vagrant in Sabah but not yet recorded from Sarawak. **HABITS AND HABITAT** Generally adaptable; can be found singly or in small groups in inland freshwater and man-made wetlands such as oxidation ponds and former tin-mining lakes/pools. Spends much time in water swimming, diving to feed or when alarmed. Can be quite aggressive during breeding season, chasing other individual males over the water's surface to protect its territory.

LEFT: *Adult*. RIGHT: *Juvenile*

Oriental Darter ■ *Anhinga melanogaster* (Burung Kosa) 85–95cm

DESCRIPTION Large, slim black bird with small head, snake-like neck and pointed yellowish bill. Plumage is generally black, with conspicuous pale streaks on wings, especially when sunning in the open. Head and neck are brown with a narrow white stripe running from gape down side of neck. Flies with neck outstretched. **DISTRIBUTION** India and Southeast Asia to the Sundas. Scattered throughout lowland Sabah and Sarawak; the

Kinabatangan wetlands in Sabah are a major stronghold for the resident population. No longer found in Peninsular Malaysia, and no records from Singapore. **HABITS AND HABITAT** Mainly restricted to remote inland freshwater swamps, lakes and rivers, and coastal forest. When swimming, its body is slightly submerged, leaving only head protruding. Often seen sunning for long periods on exposed branches with outstretched wings and tail to dry its waterlogged plumage. Roosts communally on bare trees near the riverbank.

Storm's Stork ■ *Ciconia stormi* (Burung Botak Hutan) 85cm

DESCRIPTION Medium-sized stork. Mainly black on crown, upperparts and underparts, but throat, hind-neck and tail (usually visible only in flight) are white. Bill and legs red. Very large, bright yellow eye-ring is diagnostic. **DISTRIBUTION** Malay Peninsula, Sumatra and Borneo, including Sabah and Sarawak; not in Singapore. Rare resident in lowlands, reduced everywhere but with a stronghold in the Kinabatangan wetlands, Sabah, and remnant populations in Taman Negara National Park, Panti Forest Reserve, Baram River and a few other sites. **HABITS AND HABITAT** Confined to streams, rivers and pools in dense lowland forests and peat-swamp forests. Encounters are with solitary birds or small groups feeding at edge of water or soaring on thermals. At times, can be seen perched on emergent forest trees or exposed bare branches.

Lesser Adjutant ■ *Leptoptilos javanicus* (Burung Botak Babi) 110cm

DESCRIPTION Largest stork in Malaysia. Slate-grey to black upperparts and wings, and white underparts; head and neck are largely bare with scattered tufts, showing yellowish skin and pouch, this turning brighter orange when breeding. Iris white; bill massive and greyish white or ivory coloured, as are legs. **DISTRIBUTION** India, S China and Southeast Asia to the Greater Sundas. Resident in small numbers in Peninsular Malaysia, Sabah and Sarawak; increasingly often as a vagrant in Singapore. **HABITS AND HABITAT** Generally found in coastal wetlands but can occur further inland in rice fields or grassland. Often seen on mudflats, slowly pacing in search of prey and offal, in loose groups or solitary, or soaring on thermals; in flight, is as large as a vulture but with a heavy, straight bill and projecting legs. Known to nest on tall, isolated or emergent trees.

Yellow Bittern ■ *Ixobrychus sinensis* (Burung Pucong Merah) 37cm

DESCRIPTION Plumage generally yellowish brown with lighter tones on wings. Throat and underparts white with brown streaks. Black crown and primaries (sometimes

not visible when standing); in flight, distinctly 3-coloured cream, brown and black appearance. Iris orange-yellow. Bill yellow with some black on upper mandible. Immatures heavily streaked. **DISTRIBUTION** India, China and Southeast Asia through to New Guinea. Resident in Peninsular Malaysia and Singapore, Sabah and Sarawak, the population boosted by migrants outside the breeding season. **HABITS AND HABITAT** Solitary individuals usually seen hiding or stalking prey in midst of thick vegetation and reedbeds at edge of wetlands such as rivers and canals. Very cautious, will take flight easily if approached too close. If among reeds or tall grasses, will freeze and point bill upwards to blend in when alarmed.

Female

Cinnamon Bittern ■ *Ixobrychus cinnamomeus* (Burung Pucong Bendang) 37cm

DESCRIPTION Like Yellow Bittern (above) but overall adult plumage mainly rich cinnamon on head, nape, upperparts and flanks, with a lighter tone on throat and

underparts. Bill yellow with some black on upper mandible. Iris and feet yellow. Immatures more mottled and streaked. **DISTRIBUTION** India to China and through Southeast Asia to the Greater Sundas, Philippines and Sulawesi. Resident in Peninsular Malaysia and Singapore, Sabah and Sarawak, boosted by migratory individuals outside the breeding season. **HABITS AND HABITAT** Individuals generally seen at inland wetlands such as rice fields and former tin-mining pools. Like the Yellow Bittern, is very shy and secretive, and will always seek to hide in vegetation cover.

Black Bittern ■ *Ixobrychus flavicollis* (Burung Pucong Hitam) 58cm

DESCRIPTION Dark bronzy black, browner in female (shown here), with dark bill and legs; rich buff on neck and upper breast, heavily streaked. In flight, very dark with slow, steady wingbeats, and almost as big as a night-heron. **DISTRIBUTION** India and Sri Lanka through S China to Taiwan, southwards through Southeast Asia to New Guinea, Australia and the Solomon Islands; apparently a non-breeding migrant in Peninsular Malaysia, Singapore, Sabah and Sarawak. **HABITS AND HABITAT** Mainly nocturnal, with peak activity at dawn and dusk, and seldom seen by day. Perches among tall, thick reeds and grasses at edges of ponds and rivers, where it can occasionally be spotted in torchlight from a boat. Migrants can occur in any habitat, from lowland forest to towns, but do not linger. Little information exists on behaviour and diet, because the birds are so difficult to observe.

Black-crowned Night-heron

■ *Nycticorax nycticorax* (Burung Pucong Kuak) 60cm

DESCRIPTION Handsome, small but stocky heron. Mainly ashy-grey plumage with black short bill, cap and mantle in adult, and creamy-white underside. 2 white head plumes are sometimes visible. Iris red and legs yellow, brighter when breeding. Immatures mainly scruffy-looking, streaked and mottled brown, similar to immature Little Heron (p. 20) but with pale spots rather than streaks. **DISTRIBUTION** Widespread in both New and Old worlds except Australia. Resident in Peninsular Malaysia and Singapore, Sabah and Sarawak, but localised and patchy. **HABITS AND HABITAT** The only nocturnal heron through most of the region, feeding mainly at night in shallow water. Forms large, noisy roosts and breeding colonies at man-made or natural wetlands, mostly in coastal mangroves; numbers fluctuate and colonies are displaced by habitat loss and direct persecution.

Little Heron ■ *Butorides striata* (Burung Pucong Keladi) 45cm

DESCRIPTION Commonest small heron in region. Slaty grey with a dark crown, pale face markings and pale fringes to wing feathers. Juveniles are brownish, spotted above and heavily streaked below. Bill dark, and legs varying from greenish to yellow or bright

orange depending on age and breeding status. There is much plumage variation between individuals depending on sex and geographical origins. **DISTRIBUTION** Through much of the tropics and subtropics in South America, Africa and Australasia, and in Asia extending as far N as Japan and Ussuriland. Common throughout Southeast Asia as a resident and migrant. **HABITS AND HABITAT** Generally seen singly, quite often in flight but usually stalking or standing on mud at the edge of rivers, mangroves or the sea, or even along small forest streams. Small fish, crabs and other invertebrates are the main foods.

Chinese Pond-heron

■ *Ardeola bacchus* (Burung Pucong Cina) 45–52cm

DESCRIPTION Non-breeding birds have light brown upperparts and wing coverts; brown streaks from head to chest, and white underparts and wings; bill blackish with some yellow on lower mandible; legs yellow. When breeding, head, throat, nape and breast turn a rich chestnut; mantle black; underparts remain white; feet orange. **DISTRIBUTION** Resident from India to China and Southeast Asia. Common non-breeding visitor to Peninsular Malaysia and Singapore, less common in Sabah and Sarawak. **HABITS AND HABITAT** Non-breeding birds are encountered in small numbers in a variety of wetlands, from coastal mangroves to former tin-mining pools, often solitary or in loose associations. Prior to spring migration, some individuals will assume full or partial breeding plumage. At times, birds form large communal roosts, as recorded in the Teluk Air Tawar–Kuala Muda rice scheme area on mainland Penang.

ABOVE LEFT: *Adult breeding.* LEFT: *Non-breeding*

Eastern Cattle Egret ▪ *Ardea ibis* (Bangau Kendi) 50–55cm

DESCRIPTION Small, stocky egret with a comparatively short neck and bill, and bunchy feathers beneath chin. Non-breeding birds have a yellow bill and black legs, but bill and legs turn red when breeding and cinnamon-orange patches develop on plumage of head, neck, chest and back. **DISTRIBUTION** Nearly global distribution, having aggressively expanded range over past decades. Mostly a migrant to Peninsular Malaysia, Singapore, Sabah and Sarawak, but breeding recorded at 1 site in Peninsular Malaysia; feral birds occur year-round in Singapore. Wild migrants begin to assume breeding plumage in Feb, before departure, but feral birds do so from Dec onwards. **HABITS AND HABITAT** Unlike other white egrets, prefers grasslands and pastures as well as nearby wetlands such as rice fields and former tin-mining pools. Seeks the company of cows or Water Buffalo, feeding off insects disturbed by their grazing movements.

Intermediate Egret
▪ *Ardea intermedia* (Bangau Kerbau) 65–72cm

DESCRIPTION Like Great Egret (p. 22) when not breeding, and hard to distinguish except by direct comparison, but smaller in size and has a smooth, S-shaped neck without kink. During breeding, long head and breast plumes form; bill black and lores yellow. **DISTRIBUTION** Africa, S and E Asia down to Southeast Asia and Australia. Non-breeding migrant in Peninsular Malaysia and Singapore, Sabah and Sarawak. **HABITS AND HABITAT** Shares similar wetland habitats and habits with Great Egret, including both freshwater and coastal areas, mangroves, mudflats and paddy fields. Found in loose flocks or singly, and will congregate with other white egrets to feed or breed.

Little Egret ■ *Egretta garzetta* (Bangau Kecil) 55–65cm

DESCRIPTION Slim, graceful, very active egret, slightly larger but markedly thinner-necked than stocky Eastern Cattle Egret (p. 21). Non-breeding birds sport slender black bill, legs

and feet, and yellowish lores. Breeding birds develop head, back and breast plumes. At all seasons, look for contrast between dark legs and yellow toes. **DISTRIBUTION** Africa, Europe, Asia and Australia. Mainly passage migrants and winter visitors to Malaysia and Singapore, but scattered records of breeding in Peninsular Malaysia and Sabah. **HABITS AND HABITAT** Mixes with other white egrets, mainly Great Egret and Intermediate (p. 21), in both natural and man-made wetlands. Habits are similar to those of other white egrets, but more actively chases food items, and stirs shallow water with 1 foot to disturb prey into movement.

Eastern Great Egret
■ *Ardea modesta* (Bangau Besar) 85–100cm

DESCRIPTION Largest of all white egrets in region, with an unmistakable angular kink in neck. Non-breeding adults (as shown) show bright yellow bill and lores, and black legs. When breeding, lores turn blue, bill black and long head plumes emerge; leg colour remains unchanged, or may develop red 'knees'. **DISTRIBUTION** Global distribution. Mostly migrants in Peninsular Malaysia, Singapore, Sabah and Sarawak, but small breeding populations in Sabah and Peninsular Malaysia. **HABITS AND HABITAT** Found in a wide variety of natural and man-made wetlands. Seen singly or in same-species flocks, or in the company of Little and/or Intermediate Egrets (p. 21). Mostly a stand-and-wait hunter, but follows behind tractors with other white egrets in rice fields to take advantage of ploughed-up invertebrates and frogs, and can 'helicopter' to pick prey from surface of sea. A graceful flyer with slow wingbeats and neck held in 'S' position.

Grey Heron

■ *Ardea cinerea* (Burung Pucong Seriap) 90–98cm

DESCRIPTION One of the large herons in region. Plumage is generally grey, white on head, neck and underparts, and darker on mantle and wings. Broad black eye-stripe, primaries and head plume (may not be visible at times). White central stripe down front of neck is bordered by irregular black streaks; black patch on sides of body near bend of wing. Bill and legs yellow, brighter when breeding. **DISTRIBUTION** Widely distributed across Africa, Europe, continental Asia and the Sundas. Common in Southeast Asia as a resident and migrant. Both resident and migrant in Peninsular Malaysia and Singapore, but so far recorded only as a migrant in Sabah and Sarawak. **HABITS AND HABITAT** Found along coastal and inland wetlands, usually stalking fish in shallow waters. Makes deep, guttural calls when alarmed. Forms large nesting colonies in trees, building untidy stick nests.

Purple Heron ■ *Ardea purpurea*
(Burung Pucong Serandau) 80–90cm

DESCRIPTION Almost same size as Grey Heron (above) but slimmer. Plumage ashy grey at base of neck, upper body and wings. Head and neck rufous with a black stripe from base of gape down to belly. Cap black. Bill yellow with some black on upper mandible; feet light yellow. In flight, separated from Grey Heron by dark plumage, chestnut underwings and skinny appearance. **DISTRIBUTION** Africa, Europe and Asia to the Sundas. Resident and migrant in Peninsular Malaysia and Singapore; so far, known only as a resident in Sabah and Sarawak. **HABITS AND HABITAT** Associated with brackish and freshwater wetlands. Mainly solitary, often hunting quietly for fish in the shallow waters by stalking and stabbing. Like the Grey Heron, forms large breeding colonies, sometimes with other herons such as Black-crowned Night-herons (p. 19), but often nests in thick vegetation on the ground.

White-fronted Falconet
■ *Microhierax latifrons* (Rajawali Dahi Putih) 16cm

DESCRIPTION One of the world's smallest raptors, mainly black above and white below, the white continuing in a broad patch across cheek and on forehead. Female has rufous forehead. **DISTRIBUTION** Endemic to Borneo, and there found only in Sabah. Elsewhere in Southeast Asia, Singapore and the rest of Borneo, is replaced by the similar **Black-thighed Falconet** M. *fringillarius*, which has a narrow white line behind eye curling around a black cheek patch; narrow white forehead, throat, breast and marks on tail; rufous belly and vent; juveniles have rufous tinge to pale band from cheek to brow. **HABITS AND HABITAT** Sociable, forming loose family associations in the canopy of lowland rainforest, where individuals sit separately on bare, exposed perches, visible to each other up to several hundred metres apart, sometimes exchanging perches and sharing prey. Feeds on large insects (katydids, locusts, cicadas) and small birds.

Peregrine Falcon
■ *Falco peregrinus* (Burung Falko Peregrin) 40–48cm

DESCRIPTION Large, dark falcon. Resident race F. *p. ernesti* is shown here, in which the black hood is distinctive; upperparts and tail dark grey; throat and underparts cream with dark streaks and heavy barring. Juveniles of resident race rich rufous below, but told from Oriental Hobby F. *severus* by large size, and by barring rather than streaks below. Migrants are paler, and hood is broken by pale cheeks separating dark moustache from rear of head and neck. Narrow orbital ring and legs yellow; bill yellow at base and dark towards tip. **DISTRIBUTION** Global distribution. Resident mainly near limestone cliffs in Peninsular Malaysia, Singapore, Sabah and Sarawak; migrants more widespread throughout region. **HABITS AND HABITAT** Peregrines are usually solitary or in pairs, seen at limestone outcrops in forest, and in open country where residents are much outnumbered by migrants. Both residents and migrants have been recorded using city buildings in lieu of natural cliffs. They use high vantage points to look out for prey – mainly smaller birds such as pigeons and waders, sometimes bats – and hunt by diving and knocking them over in flight.

Osprey
■ *Pandion haliaetus* (Lang Tiram) 54cm

DESCRIPTION Medium-sized raptor with brown upperparts and tail. Head white with prominent brown/black eye-stripe. Short crest, but not always visible. Underparts and legs white with brown patches on chest. Iris yellow. **DISTRIBUTION** Globally distributed, breeding in temperate and subtropical regions, and wintering in tropical areas. Non-breeding visitor at low elevations in Peninsular Malaysia, Singapore, Sabah and Sarawak. **HABITS AND HABITAT** Seen along sea coasts, and on inland natural and man-made wetlands such as former tin-mining areas and dams, usually in flight. Feeds primarily on fishes, which it catches by diving spectacularly into water from the air before emerging and taking off from surface with prey in talons. Recorded in all months, and migrants may arrive from both N and S hemispheres.

Black Baza ■ *Aviceda leuphotes* (Lang Baza Hitam) 32cm

DESCRIPTION Beautiful black and white raptor, typically seen in flight with rounded, butterfly-like wings. Black, with large white wing-patches; white below with broad black breast-band, and belly often strongly and widely barred blackish chestnut and white.

DISTRIBUTION From India, the Himalayas and Nepal through S China to Hainan and continental Southeast Asia as far as 14°N. Migrant to Southeast Asia at least to Sumatra and Java, including Peninsular Malaysia and Singapore; not recorded from Sabah or Sarawak. **HABITS AND HABITAT** Small parties soar on broad, rounded wings, showing typical pied appearance. They occur over wooded habitats of all kinds in lowland plains, including rubber and oil-palm plantations, secondary woodland and forest, flying down to take insects from foliage and also resting in canopy. In the S part of Peninsular Malaysia migrants arrive in last week of Oct, and heaviest northward passage is in late Mar.

Sunda Honey-buzzard ■ *Pernis ptilorhyncus* (Lang Lebah) 55–65cm

DESCRIPTION Medium-large raptor with variable individual plumages. In flight, shows a longish, proportionately small, chicken-like head, a long, square-cut tail with 2–3 dark bands, and long wings with numerous dark bands on undersurface. Upperparts brown to chocolate or blackish; underparts from cinnamon to white, with highly variable amount

of streaks and barring. **DISTRIBUTION** Now considered two species: **Sunda** *Pernis ptilorhyncus* resident in Malay Peninsula, Sumatra, Borneo (Sabah, Sarawak) and Java (and disperses to Singapore), while **Oriental** *Pernis ruficollis* is migrant from N, E and S Asia to same area and beyond. **HABITS AND HABITAT** Seen in large numbers at key migration points such as Tanjung Tuan (Port Dickson) and Taiping. Prefers forests (the resident race) as well as plantations and other wooded areas, and feeds on honey-bee and wasp larvae by raiding their nests. However, also said to take small vertebrate prey and chickens in

Resident P. ptilorhyncus *(left)*; *migrant* P. ruficollis *(right)* villages opportunistically.

Bat Hawk

■ *Macheiramphus alcinus* (Lang Malam) 45cm

DESCRIPTION Mainly black, with small crest, light yellow eye, and variable amount of white on throat and upper breast, this marred by some black streaks. Notable for rapid flight, with powerful, deliberate strokes of its long, narrow-tipped wings. **DISTRIBUTION** Tropical Africa and Madagascar; and Malay Peninsula to Sumatra, Borneo, Sulawesi and New Guinea. Resident in Peninsular Malaysia, Sabah and Sarawak; scarce non-breeding dispersant to Singapore. **HABITS AND HABITAT** Typically seen in the late afternoon and evening, in forest especially near limestone cliffs with caves, where bats emerge and are hunted; also over secondary woodland and open country. Bat Hawks can overhaul most bats in direct flight, or grab one from the edge of a flock, using the foot to catch the bat and then swallowing it in flight. Its big stick nests have been found a number of times, pairs using the same nest site repeatedly.

Black-shouldered Kite ■ *Elanus caeruleus* (Lang Tikus) 32cm

DESCRIPTION Elegant, fairly small raptor with light grey crown, darker grey upperparts and black primaries. Face and underparts white to cream or pearly grey. Bill black, iris red, eyebrows black and legs yellow.

DISTRIBUTION Africa, S Eurasia, India and S China through Southeast Asia to New Guinea. Resident at low elevations in Peninsular Malaysia, Singapore, Sabah and Sarawak.

HABITS AND HABITAT A raptor of open country and agricultural areas, such as rice fields, oil-palm plantations and grassland. Often seen hovering in search of rodents during daytime, but also uses utility poles as a hunting perch.

Brahminy Kite ■ *Haliastur indus* (Lang Merah) 45cm

DESCRIPTION Medium-sized raptor. Adults sport a dark-streaked white head, nape and chest, and a chestnut-brown body; legs yellow; bill light yellow. Immatures are mostly brown with streaked breast.

DISTRIBUTION India, S China and throughout Southeast Asia to Australia. Common resident at low elevations in Malaysia and Singapore.

HABITS AND HABITAT One of the common raptors along coastal areas, but can also be found further inland, above rice fields, former tin-mining wetlands and cities/towns with urban wetlands. Usually seen soaring most of the time over its preferred habitats looking for live prey or scavenging. Encounters range from solitary birds to large groups.

White-bellied Fish-eagle

■ *Haliaeetus leucogaster* (Lang Siput) 70cm

DESCRIPTION Large raptor with white head, underparts and tail; upperparts and wings black; bill grey, iris brown/black, legs light yellow. Tail wedged, visible in flight. Immatures in various stages of mottled brown. **DISTRIBUTION** India and Southeast Asia to Australia. Resident in Malaysia and Singapore. **HABITS AND HABITAT** Common raptor frequenting coastal wetlands and even large inland wetlands such as former tin-mining areas and dams. Feeds mainly on fish. Hunts by soaring and circling over water bodies before swooping down swiftly to catch fishes near the water surface. Builds a large nest on tall canopy trees or man-made structures, e.g. telecommunication towers. Pairs are very vocal.

Crested Serpent-eagle ■ *Spilornis cheela* (Lang Kuik) 54cm

DESCRIPTION Medium-sized, generally brownish raptor. Upperparts greyish brown; underparts lighter brown with small white spots on breast, belly and top of shoulder. Head

darker brown (almost black in some individuals). Cere, iris and feet yellow. Bill grey. Short crest (not always visible when perched). Immatures mottled brown. **DISTRIBUTION** India, S China and Southeast Asia to Greater Sundas. Resident in Malaysia and Singapore. **HABITS AND HABITAT** Usually solitary, often heard or seen in various habitat types, e.g. mangroves, forests, residential areas (with nearby forested areas) and oil-palm plantations. At times, seen perched on vantage points. Diet consists mainly of snakes.

Grey-faced Buzzard ■ *Butastur indicus* (Lang Belalang) 45cm

DESCRIPTION Fairly small raptor when perched, with distinctive face pattern, black streak down centre of throat and dark bars on tail; chocolate breast with bars on flanks and belly; staring yellow eye. In flight, shows a chequered cinnamon panel on inner primaries and secondaries, and wings appear fairly long and narrow when soaring. **DISTRIBUTION** Cool temperate NE Asia, from Amurland to Japan and N China, migrating S to Java, Bali and the Lesser Sundas. Migrant in Peninsular Malaysia, Singapore and, more often, Sabah and Sarawak. **HABITS AND HABITAT** Most often seen on migration, predominantly in Oct and Mar, in open country with scattered trees and in agricultural land. Feeds on lizards, insects and, perhaps, rats.

Blyth's Hawk-eagle ■ *Nisaetus alboniger* (Lang Hantu) 52–58cm

DESCRIPTION Medium-sized raptor with black upperparts and prominent crest visible when perched; underparts white, marked with black vertical streaks on chest and horizontal barring on belly; tail black with broad white band; feet yellow. Sub-adult light sandy buff on head and breast, darker brown on back and wings, with 3 or 4 dark bars across tail. **DISTRIBUTION** From *c.* 10°N through the Malay Peninsula, Sumatra and Borneo. Resident in hills up to 1,900m in Peninsular Malaysia, Sabah and Sarawak; recorded from Singapore. **HABITS AND HABITAT** Usually seen soaring over the forest canopy, in lowlands and montane forest, or waiting on a high branch in search of prey, which includes lizards, bats and other small mammals. Builds large nest in a tall tree, at the point where several main boughs diverge to form tree crown and where flight access is easy.

Changeable Hawk-eagle

▪ *Nisaetus limnaeetus* (Lang Iindik) 60–75cm

DESCRIPTION Medium-sized raptor with 2 morphs. Dark-morph individuals are blackish brown in plumage with yellow feet; dark terminal band on tail visible in flight. Pale-morph birds are generally dark brown on upperparts with whitish underparts and head, marked with streaks on chest and belly; in flight, shows 4 or 5 dark bands across tail (up to 7 in juveniles). At times, small crest is visible in both when perched. **DISTRIBUTION** India, through Southeast Asia to the Sundas. Resident at low elevations in Peninsular Malaysia, Singapore, Sabah and Sarawak. **HABITS AND HABITAT** Pale morph is more commonly encountered than dark morph. A raptor of wooded country and forest (ranging from disturbed to primary forest), or of forest patches in a matrix of agriculture and settlements. Usually seen soaring over canopy, or heard calling a 2- or 3-note *whe Wheet!*

Wallace's Hawk-eagle ▪ *Nisaetus nanus* (Lang Hantu Kasturi) 45cm

DESCRIPTION Like a sub-adult Blyth's Hawk-eagle (p. 29), but slightly smaller, with more uniform dark upperparts and 3 (not 2) dark bands across tail. Adult Blyth's is more

black and white in appearance, whereas Wallace's has brown and rufous plumage. **DISTRIBUTION** From *c.* 12°N in Thailand through Malay Peninsula, Sumatra and Borneo; resident in Peninsular Malaysia, Sabah and Sarawak, but not recorded from Singapore. **HABITS AND HABITAT** Tall forest in level lowlands, including forest edges alongside rivers and disturbed areas, where birds perch in the lower part of the tree crown and keep a lookout for prey. Known food includes lizards, but there are few records. Builds a nest of sticks in a main fork of a tall canopy or emergent tree, and produces a single chick.

Slaty-breasted Rail ▪ *Lewinia striata* (Burung Sintar Biasa) 25cm

DESCRIPTION Slaty-grey face, throat, chest and underparts; chestnut crown; light brown upperparts and wings; white barring on belly and upperparts; feet greyish; bill red.
DISTRIBUTION India, S China and through Southeast Asia to the Sundas and the Philippines. Resident at low elevations in Peninsular Malaysia, Sabah, Sarawak aand Singapore. **HABITS AND HABITAT** Found in inland freshwater swamps, flooded rice fields, reedbeds and grasslands. Generally shy and keeps to thick vegetation, affording brief glimpses of itself as it appears at edges or small openings; remains solitary except when accompanied by chicks.

White-breasted Waterhen
▪ *Amaurornis phoenicurus* (Burung Ruak-ruak) 33cm

DESCRIPTION White face and breast, merging to rufous beneath tail, and dark back and wings make species unmistakable. Sexes are alike. **DISTRIBUTION** India, S China and Southeast Asia. Resident throughout Malaysia and Singapore, the populations there being augmented by migrants and winter visitors from the temperate zone. **HABITS AND HABITAT** Commonly seen in rank vegetation, overgrown drains and along roadsides in rural areas, sometimes flying up when disturbed. Adults may be accompanied by several half-grown fluffy black chicks, the pale breast plumage gradually appearing as they grow. Monotonous single piping note endlessly repeated, or a chorus of grating and gurgling notes in which both the male and female participate, competing with neighbouring pairs.

Watercock ▪ *Gallicrex cinerea* (Burung Ayam-ayam) 42cm

DESCRIPTION A bulky rail; the female and non-breeding male predominantly mottled brown, buff and black, with light yellow legs and bill. Breeding males are more

attractive, with prominent red frontal shield adjoining bill; black head, neck and chest; red legs. **DISTRIBUTION** India, China, Southeast Asia and the Philippines. Predominantly non-breeding winter visitor and passage migrant in Peninsular Malaysia and Singapore, Sabah and Sarawak, with scarce breeding records. **HABITS AND HABITAT** Prefers natural and man-made freshwater wetlands. Usually solitary and moves about under thick vegetation, affording observers brief glimpses at small clearings. Possesses nocturnal habits.

Female

Purple Swamphen ▪ *Porphyrio porphyrio* (Burung Pangling) 42cm

DESCRIPTION Big, gaudy rail with deep purplish-blue plumage; red bill, forehead and legs; white below tail. In poor light it can look merely blackish at a distance. Sexes

are alike, juveniles darker and duller. **DISTRIBUTION** Resident across much of Africa, the Mediterranean and Middle East, through S Asia to Australasia and Oceania, including Peninsular Malaysia, Singapore and Borneo; has recently spread and become commoner in Sabah and Sarawak. **HABITS AND HABITAT** Occasionally seen in 1s or 2s in swampy habitat, where it feeds on succulent water plants, holding material in 1 foot and slicing it with the scissor-like bill. Nest is a bowl of piled-up weeds in dense vegetation in the marsh. A variety of loud braying, chuckling and clattering notes may be heard in the morning and evening, but birds are secretive, falling silent and retreating into vegetation when approached.

Common Moorhen ▪ *Gallinula chloropus* (Tiong Air) 32cm

DESCRIPTION Dusky-black waterbird with a narrow white band along flanks and white beneath cocked-up tail. Prominent red frontal shield and bill with yellow tip.

DISTRIBUTION Almost worldwide distribution in temperate and tropical zones, except Australia and New Zealand. Resident in Peninsular Malaysia and Singapore, Sabah and Sarawak, the population having expanded during the past 3 decades; also supplemented everywhere by migrants from c. Oct–Mar. **HABITS AND HABITAT** Favours former tin-mining wetlands, flooded rice fields and canals. Often seen in small, loose groups, family parties or individually. Spends much time around water, feeding on surface vegetation and insects.

Masked Finfoot ▪ *Heliopais personatus* (Burung Pedendang) 54cm

DESCRIPTION Both sexes are generally olive-brown but slightly darker on upperparts; iris yellow; bill large and yellow; feet greenish yellow and webbed. Adult male has a black throat and face with a white line from back of eye. Female has a white throat.

DISTRIBUTION India, Bangladesh and Southeast Asia. Non-breeding winter visitor at low elevations in Peninsular Malaysia. Vagrant in Singapore and Sabah, and recently also recorded in Sarawak in a protected area. **HABITS AND HABITAT** Encounters are limited to individuals in the species' preferred habitats of forested waterways, former tin-mining wetlands and coastal mangroves. Very shy and secretive. Swims with head bobbing and usually keeps close to water's edge and vegetation.

Female

Pacific Golden Plover

■ *Pluvialis fulva* (Burung Rapang Keriyut) 25cm

DESCRIPTION Non-breeding plumage is brown, spangled with gold above and with paler buffy face, neck and underparts. After arrival and before departure, many are in part breeding plumage (as shown), with brighter upperparts and patchy black face and underparts separated by a white line from brow to neck and flanks. **DISTRIBUTION** Resident in NE Siberia and W Alaska, migrating to E Africa, and S and Southeast Asia as far as New Guinea, Australia and New Zealand. Migrant throughout coastal lowlands of Peninsular Malaysia, Sabah, Sarawak and Singapore. **HABITS AND HABITAT** Loose flocks settle on open mud or short grass, standing motionless and well camouflaged, or foraging for small bivalves, snails and worms. Takes off with a piping *kieu-wik*, especially vocal at night.

Black-winged Stilt ■ *Himantopus himantopus* (Burung Stilt) 38cm

DESCRIPTION Slim and elegant, with long pink legs and a slender bill. Head, neck and underparts are white with variable dusky markings on crown, face and neck; wings

Adult (left) and juveniles (right)

are black in adults, dusky in juveniles. Migrants from Australasia, sometimes separated as White-headed Stilt *H. leucocephalus*, have a whiter face and crown, and a black patch on hind-neck. **DISTRIBUTION** Resident through most of the temperate Old and New worlds, migrating to the tropics; migrant to coastal lowlands and fresh waters of Peninsular Malaysia (where there are increasing numbers of nestings), Singapore, Sabah and Sarawak. **HABITS AND HABITAT** Small flocks occur in wet rice fields, open marshes and, occasionally, near the sea. The few nests found have been on muddy banks of lakes or on wet vegetation.

Little Ringed Plover
▪ *Charadrius dubius* (Burung Rapang Biji Nangka) 17cm

DESCRIPTION Small plover with a pale forehead, throat and collar, a complete dark breast-band and white underparts. Small bill is all dark, ring of skin around eye is pale, and legs are pale olive to pink. Coming into breeding plumage, contrasts are heightened, black face mask is distinct, and eye-ring is a clear yellow. **DISTRIBUTION** Resident through sub-Arctic and temperate Eurasia to the Middle East, Sri Lanka, the Philippines and New Guinea; migrates S to N Africa, China, Southeast Asia and Australia. Migrant in Peninsular Malaysia, Singapore, Sarawak and Sabah. **HABITS AND HABITAT** Primarily in freshwater habitats near temporary pools, on short grass and open ground, mud, ploughed farmland and rice fields; occasionally on intertidal mud. Birds often wander separately over the feeding habitat, flying up to form a flock when disturbed.

Winter

Siberian Plover ▪ *Anarhynchus mongolus* (Burung Rapang Mongolia) 20cm

DESCRIPTION Moderate-sized plover. Non-breeding plumage is grey-brown and white; forehead, brow, throat and narrow collar white, separated from white underparts

by an incomplete grey-brown breast-band. Legs and bill dark. Similar **Greater Sand-plover** *C. leschenaultii* is larger, with a heavier bill, more complete breast-band and paler legs. Many birds show fragmentary breeding plumage with a chestnut breast-band. DISTRIBUTION Resident in sub-Arctic Russia and Siberia S to China; migrates S to coasts of Indian Ocean, Southeast Asia and Australasia, including all of Malaysia and Singapore, where it is a passage migrant and non-breeding visitor. **HABITS AND HABITAT** An abundant migrant in small flocks, foraging on intertidal mudflats, where it seeks worms and small bivalves in the soft mud.

Winter

Greater Painted-snipe ▪ *Rostratula benghalensis* (Burung Meragi) 24cm

DESCRIPTION Female deep chestnut on head, neck and breast, this sharply set off from white around eye and on underparts; buff central crown-stripe, and brown back and wings beautifully mottled and spotted. Male is duller, with grey-brown on breast and buff around eye. Legs greenish and slightly downcurved bill orange in both sexes. **DISTRIBUTION** Africa, Madagascar and Indian sub-continent through S China to Japan, and Southeast Asia through the Lesser Sundas to Australia. Resident in Peninsular Malaysia, Singapore, Sabah and Sarawak. **HABITS AND HABITAT** This species has reversed sexual dimorphism, the duller male incubating the eggs and tending the chicks. Greater Painted-snipe are very secretive and hard to see, creeping among dense foliage such as water hyacinths and reeds in swamps and rice fields. They tend to be more active at dusk, and may be commoner than they seem.

Male (left) and female (right)

Black-tailed Godwit ▪ *Limosa limosa* (Burung Kedidi Ekor Hitam) 40cm

DESCRIPTION Tall wader, here with 3 Common Redshank. Non-breeding plumage is plain grey above and off-white below; blackish legs and black tip to pink bill. In flight, shows a white band across rump and base of tail, and white wing bar. In breeding plumage, is largely rufous on head, upperparts and breast, with a barred black and whitish belly and flanks. **DISTRIBUTION** Sub-Arctic and temperate Eurasia from Iceland to Siberia, migrating S to N Africa and SW, S and Southeast Asia as far as New Guinea and Australasia. Non-breeding migrant in coastal lowlands throughout Malaysia and Singapore. **HABITS AND HABITAT** Found on intertidal mudflats and around pools in open mangroves, roosting in the lower branches of mangrove trees at high tide; small to very large flocks forage for bivalves and worms in the mud.

Whimbrel ■ *Numenius phaeopus* (Burung Kedidi Pisau Raut) 44cm

DESCRIPTION Large wader, mottled, spotted and barred with brown and buff; best recognised by dark lateral stripes on crown with central pale line, and long, curved bill.

Most show white rump in flight, but some are from dark-rumped population. The scarcer Eurasian Curlew *N. arquata* is also white-rumped but bigger, and has no bold crown-stripes and a much longer bill. **DISTRIBUTION** Resident in Arctic and sub-Arctic Eurasia and parts of North America; migrant to South America, Africa, and S and Southeast Asia as far as Australasia, including Peninsular Malaysia, Sabah, Sarawak and Singapore. **HABITS AND HABITAT** Typically on coastal mudflats fronting mangroves, sometimes in big flocks, probing mud or wet sand for worms. Call is a clear musical trill, often uttered when taking flight.

Terek Sandpiper ■ *Xenus cinereus* (Burung Kedidi Sereng) 25cm

DESCRIPTION Small wader with pale grey-brown upperparts, slightly darker carpal-patch on wing, and pale brow. Orange-yellow legs and orange base to upturned bill. In flight,

rump and tail are same colour as back, but secondaries show a white trailing edge to wing. **DISTRIBUTION** Resident in Arctic and cool temperate Eurasia, from Scandinavia to the Amur; migrates to South Africa and the borders of the Indian Ocean as far as S and Southeast Asia, Australia and New Zealand. Migrant in moderate numbers to Peninsular Malaysia, Singapore, Sabah and Sarawak. **HABITS AND HABITAT** On sandy beaches and mudflats, foraging for invertebrates in soft intertidal mud. Call is a ringing *kleet-kleet* when flushed, with emphasis on the 1st note.

Common Sandpiper ■ *Actitis hypoleucos* (Burung Kedidi Biasa) 20cm

DESCRIPTION Small wader with pale grey-brown upperparts; white below, with distinctive brown patch on each side of upper breast; pale brow and eye-ring. Dark bill and olive legs. In flight, shows a white wing bar and white sides to narrow brown rump.

DISTRIBUTION Resident across N Eurasia from W Europe to Japan and S to Iran, the Himalayas and China; migrates S to Africa, and S and Southeast Asia as far as the Philippines and Australia, rarely to W Pacific islands and New Zealand. Migrant in Peninsular Malaysia, Singapore, Sabah and Sarawak, but with sightings in all months. **HABITS AND HABITAT** Abundant but thinly distributed in many habitats, including coasts, wet rice fields, rivers, ditches and even concrete-lined drains in towns. Usually alone or in pairs, seen teetering along with a bobbing tail. Makes a shrill, piping call when flushed, as it flies low over water on bowed wings.

Common Greenshank ■ *Tringa nebularia* (Burung Kedidi Kaki Hijau) 35cm

DESCRIPTION Fairly large, slim wader, light grey above and whitish below, with a long, dark bill and greenish legs. In flight, shows a wedge of white from rump to back,

Non-breeding

but no white wing bar. In breeding plumage, face, upper breast and back are more spotted and mottled. **DISTRIBUTION** Resident across N temperate Eurasia from Europe to the Amur; migrates S to tropical Africa, and SW, S and Southeast Asia as far as the Philippines, Australia and New Zealand. Migrant throughout Peninsular Malaysia, Singapore, Sabah and Sarawak. **HABITS AND HABITAT** A common migrant on mudflats; also occurs in small numbers inland on wet rice fields or marshes. Numbers increase through Oct and decline markedly in Mar, but like many wader species 1 or 2 birds can be found in almost any month.

Wood Sandpiper ■ *Tringa glareola* (Burung Kedidi Sawah) 23cm

DESCRIPTION Small, lively wader, dark ashy grey with a distinctive pale brow and mottled back and wings. In flight, shows a barred tail and squared-off white rump, this not extending in a wedge up back (cf. Common Redshank, below). Dark bill and yellowish legs. **DISTRIBUTION** Resident in Arctic and cool temperate Eurasia, from Europe to Kamchatka and NW China; migrates to Africa, and SW, S and Southeast Asia as far as the Philippines, New Guinea and Australia. Non-breeding migrant throughout Peninsular Malaysia, Singapore, Sabah and Sarawak. **HABITS AND HABITAT** Very common in freshwater marshes, wet rice fields and mangroves, and where swamp forest has been newly cleared. Mixes with many other waders, but does not necessarily form cohesive flocks.

Common Redshank ■ *Tringa totanus* (Burung Kedidi Kaki Merah) 28cm

DESCRIPTION Lightly mottled grey-brown above, paler below, and with a dark line before eye; red legs and red-based bill. In flight, shows a wedge of white from rump to back, and broad white panels on secondaries and inner primaries. **DISTRIBUTION**

Resident through N temperate Eurasia to the Himalayas and W China; migrates S to tropical Africa, and SW, S and Southeast Asia as far as the Philippines and NW Australia. Non-breeding migrant in Peninsular Malaysia, Sabah, Sarawak and Singapore. **HABITS AND HABITAT** An abundant wader of mudflats and muddy sand-flats, found usually in small or sometimes large flocks, these taking off when disturbed with a whistling *teu, teu-teu-teu*. Forages for crabs, worms and small molluscs.

Curlew Sandpiper
■ *Calidris ferruginea* (Burung Kedidi Merah) 20cm

DESCRIPTION Greyish above with pale brow, dark legs and dark, downcurved bill. In flight, white rump contrasts with dark tail, and edges of wing coverts form a narrow bar. Many birds begin to assume breeding plumage before migrating, with variable amounts of chestnut on breast, and chestnut and black chequers above. **DISTRIBUTION** Resident along the Arctic Ocean coast from the Yenisei in Siberia to NW Alaska; migrates S to Africa and the borders of the Indian Ocean as far as S and Southeast Asia, the Philippines, New Guinea, Australia and, rarely, Micronesia and New Zealand. **HABITS AND HABITAT** The upright stance and slightly elongated, curved bill are useful characters for picking out this species in mixed flocks on mudflats, where it takes mostly worms and some molluscs. It roosts on the mud, in mangrove clearings or on mangrove trees.

Ruddy Turnstone ■ *Arenaria interpres* (Burung Kedidi Kerikil) 23cm

DESCRIPTION Distinctive particoloured appearance even in non-breeding plumage, with short legs, short bill and abrupt forehead; dark brown head, upperparts and breast-band. In

flight, shows a white back, separate white base of tail, and 2 white wing bars. Breeding birds have a black and white face pattern, and rufous back and wings. **DISTRIBUTION** Circumpolar on Arctic Ocean coasts and islands, migrating S to the Americas, Africa and the borders of the Indian Ocean as far as S and Southeast Asia, the Philippines, New Guinea, Oceania, Australia and New Zealand. **HABITS AND HABITAT** Common but not usually in big flocks. Searches for molluscs on mud and sand, and between rocks, often locating them by flipping over

Intermediate plumage between breeding and non-breeding   pebbles and debris with its bill.

Little Tern ▪ *Sternula albifrons* (Burung Camar Kecil) 22cm

DESCRIPTION Small-sized, pale tern. Breeding birds have black crown, nape and eye-stripe, with white forehead; feet and bill yellow with black tip. Non-breeding birds sport black eye-stripe, white crown and lores, black bill and feet. Tail slightly forked. **DISTRIBUTION** Found in coastal temperate and tropical waters. Malaysia and Singapore have both resident and migrant populations. **HABITS AND HABITAT** Normally encountered in pairs or small, loose groups along coastal areas and estuaries, but also occasionally recorded from inland reservoirs and large rivers. Hunts for small fishes by hovering and diving.

White-winged Tern ▪ *Chlidonias leucopterus* (Burung Camar Bahu Putih) 25cm

DESCRIPTION Breeding birds have a black head, chest, underparts and upperparts, with a red bill. Wings grey and tail white. Non-breeding birds are similar to Whiskered Tern (p. 42), white crown and black dot behind ear coverts. **DISTRIBUTION** Breeds in Europe and temperate regions of Asia, wintering S to Africa, Southeast Asia and Australia. Passage migrant and winter visitor to Malaysia and Singapore. **HABITS AND HABITAT** Probably the most common migratory tern to the region's shores, frequenting coastal areas, rivers and inland wetlands such as rice fields, former tin-mining lakes/pools and reservoirs. Usually found in small groups, skimming over the water's surface to feed. At times, seen following ploughing tractors in rice fields in the company of white egrets. Perches on wooden poles and utility wires along the coast.

Winter

Whiskered Tern ▪ *Chlidonias hybrida* (Burung Camar Tasik) 27cm

DESCRIPTION Breeding birds sport a red bill and legs, black crown and nape, and grey underparts and mantle. Non-breeding birds are generally greyish, with white forehead and

sides of head, and black bill and streak behind eye. **DISTRIBUTION** Breeds in S Africa, S Europe, temperate Asia, Southeast Asia and Australia. Passage migrant and winter visitor to Malaysia and Singapore. **HABITS AND HABITAT** Similar to White-winged Tern (p. 41), and also commonly encountered on the region's coasts and inland wetlands. Congregates in small to large groups. Hunts for food by making shallow plunges or skimming low over water.

Winter

Black-naped Tern ▪ *Sterna sumatrana* (Burung Camar Sumatera) 30cm

DESCRIPTION Elegant tern with predominantly white plumage except for black nape;

deeply forked tail. Narrow black bill. **DISTRIBUTION** Found in tropical waters of the Indian and Pacific oceans. Breeds on small rocky outcrops and islets in the Melaka Straits and South China Sea. More commonly encountered along the E coast of Peninsular Malaysia. **HABITS AND HABITAT** Congregates in small groups and, at times, with other terns. Does not move inland, unlike some other terns.

Spotted Dove ■ *Spilopelia chinensis* (Tekukur Biasa) 30cm

DESCRIPTION Overall light brown, with a slight pinkish hue and darker upperparts. Black patch with white spots across neck is diagnostic; white vent; light yellow iris; red feet. **DISTRIBUTION** Found across India and China to Southeast Asia. Common resident in Malaysia and Singapore. **HABITS AND HABITAT** Inhabits open country, scrub, plantations, gardens and villages. Usually solitary or in pairs, feeding on the ground or engaged in courtship. Takes off when disturbed. Sometimes kept as a cagebird.

Little Cuckoo-dove ■ *Macropygia ruficeps* (Tekukur Api) 30cm

DESCRIPTION Beautiful highland dove with rich chestnut plumage that has a slightly darker tone on wings and mantle; barrings on upperparts and black mottling on chest. Iris white. Adult males have a glossy green and lilac nape, this lacking in the females. **DISTRIBUTION** Confined to Southeast Asia. Common hill and montane resident in Malaysia. Not recorded in Singapore. **HABITS AND HABITAT** Regularly encountered at hill stations such as Fraser's Hill in pairs or small flocks. A fast, powerful flyer above the canopy, capable of covering great distances in search of food. Generally shy, but its presence is often betrayed by its calls, an incessant, quick *wup wup wup wup....*

Female

Asian Emerald Dove
■ *Chalcophaps indica* (Punai Tanah) 25cm

DESCRIPTION Handsome dove with diagnostic emerald-green wings and black primaries. Crown greyish and underparts dull maroon. Bill red, legs greyish pink. Males are brighter than females, with a whiter brow and 3 bold, dark and light grey bands across back and rump. **DISTRIBUTION** Resident in India, across S China to Southeast Asia and Australia. Resident in Peninsular Malaysia, Singapore, Sabah and Sarawak. **HABITS AND HABITAT** Keeps to the ground in forests, forest edges, mangroves and plantations. Usually solitary, feeding on fallen seeds and grubs. Flight is usually low between the trees, on rapid wingbeats. Call is a soft, low *tick-Whooo*, repeated monotonously; calling birds may remain at a site for several weeks before moving on, and long-distance movements have been recorded.

Zebra Dove ■ *Geopelia striata* (Merbok Balam) 20cm

DESCRIPTION Overall brownish plumage with a slightly darker shade on head, mantle and wings; heavily barred on nape, sides and flanks; forehead and chin slightly ashy.

Eye-ring light blue, legs pinkish. **DISTRIBUTION** Myanmar and Southeast Asia to Australia. Resident in Malaysia and Singapore. **HABITS AND HABITAT** Commonly seen in open country, plantations and gardens in pairs or solitary. Forages mainly on the ground. Very vocal, and can be surprisingly tame and confiding at times. Call is an initial flourish followed by a series of single notes, *Coodle-oo coo coo coo…*; individual variation in the call is the basis for popular bird-singing competitions.

Little Green Pigeon
■ *Treron olax* (Punai Daun) 20cm

DESCRIPTION The smallest of the *Treron* pigeons. Both sexes have green plumage, yellow wing bars, black terminal band on tail, yellow eye-ring and red legs. Adult males have a greyish head and nape, prominent orange patch on chest, and maroon mantle and wing coverts. Females have a light greyish cap and green wing coverts. **DISTRIBUTION** Malay Peninsula to the Greater Sundas. Resident in Malaysia and Singapore. **HABITS AND HABITAT** Favours the canopy and/or middle storey of forests and forest edges. Can be seen at times feeding on fruiting fig trees and shrubs with other *Treron* pigeons.

Pink-necked Green Pigeon ■ *Treron vernans* (Punai Kericau) 27cm

DESCRIPTION Adult males have a grey head and throat, subtle pink on neck and breast, orange chest and light green belly; mantle and upperparts green. Females are mainly green. Both have yellow wing bars and red legs, and a pinkish/reddish eye-ring. **DISTRIBUTION** Malay Peninsula, Borneo, the Philippines and the Indonesian islands of Sumatra, Java, Bali and Lesser Sundas. Resident in Malaysia and Singapore. **HABITS AND HABITAT** Possibly the most common of the green-pigeons, found at low elevations from coastal mangroves and open country to forest edges and secondary forests. At times, will visit wooded urban gardens like the Lake Gardens in Kuala Lumpur. Usually in pairs but will congregate in large numbers to feed in fruiting trees and shrubs. Tends to stay in tree crowns.

Thick-billed Green Pigeon ■ *Treron curvirostra* (Punai Lengguak) 27cm

DESCRIPTION Both sexes have overall olive-green plumage, with yellow wing bars, light green eye-ring, thick bill with maroon at base, and red legs. Adult males have maroon

mantle and wing coverts; vent cinnamon. Females have darker olive-green wing coverts. **DISTRIBUTION** India, Nepal and Southeast Asia. Resident in Malaysia and Singapore. **HABITS AND HABITAT** Frequents mangroves, well-wooded gardens, forest edges and forests; usually seen in the canopy or middle storey. Often feeds in large parties (sometimes 50 birds) in fruiting fig trees, and at times with other frugivorous birds.

Male (left) and female (right)

Mountain Imperial Pigeon ■ *Ducula badia* (Pergam Gunung) 46cm

DESCRIPTION Large, sombre pigeon with light grey plumage and whitish or light grey throat; mantle, wings and tail dark brown with bronze iridescence; legs pink.

DISTRIBUTION India, S China, Southeast Asia, Sumatra, Java and Borneo. Resident in Peninsular Malaysia, Sabah and Sarawak, but absent in Singapore. **HABITS AND HABITAT** One of the pigeons commonly encountered in montane forests above 900m, where it remains in tree crowns and can be inconspicuous until it moves or calls. 1 or 2 are often seen feeding together with other frugivores at fruiting fig trees. Call is a deep, resonating *whoo-Whoomp*.

Long-tailed Parakeet ▪ *Psittacula*
longicauda (Bayan Nuri) Male 42cm; female 30cm

DESCRIPTION Moderately sized, bright green parrot, with a dark crown, reddish face and black throat, the wings and tail bluer than the back. All the markings and colours are brighter and more distinct in the male, which has a red rather than black bill, and a longer tail. **DISTRIBUTION** Resident in Peninsular Malaysia, Sumatra and Borneo, including Sabah and Sarawak, and extending to the Andamans and S Thailand and Myanmar. **HABITS AND HABITAT** Rocketing flocks pass across the canopy of lowland forest, or even parks and gardens where they are not persecuted, seeking trees that bear small, hard fruits. Evening flocks gathering to roost can contain hundreds of birds, all screeching as they fly. Pairs nest in tree-holes in dead standing timber, or enlarge crevices in live trees.

Blue-crowned Hanging Parrot
▪ *Loriculus galgulus* (Burung Serindit) 14cm

DESCRIPTION Smallest parrot in the region, generally green in both sexes. Adult males have red breast and rump, yellow/orange patch on mantle, and small blue crown.

Females lack the blue crown and red breast-patch. **DISTRIBUTION** Malay Peninsula, Sumatra and Borneo. Resident in Malaysia and Singapore. **HABITS AND HABITAT** Typically confined to the tree canopy in forests, forest edges and wooded gardens. Seldom descends low except to feed on small fruit and flower buds. Interestingly, will hang upside down like a bat when roosting. Often calls in flight, a high-pitched single note, soft but carrying.

Bornean Ground-cuckoo
▪ *Carpococcyx radiatus* (Burung Butbut Tanah) 60cm

DESCRIPTION Large, lanky ground bird, with a black head, back, wings and tail glossed with green and purple; neck and breast grey, shading to lightly barred flanks and belly.

Heavy bill, bare facial skin and legs are light green. Juveniles are browner and less glossy, with the breast light rufous. **DISTRIBUTION** Endemic to Borneo, where it is found in both Sabah and Sarawak. **HABITS AND HABITAT** Difficult to detect as it moves quietly among the litter or on low branches, either alone or in pairs, seeking beetles and other invertebrates. Reputed to follow migrating Bearded Pigs *Sus barbatus*, presumably to catch animals disturbed by them or to pick up scattered fragments of fruit. Utters long series of quick, deep coos, followed by long series of two-note, guttural *poop-poo* notes.

Asian Koel ▪ *Eudynamys scolopaceus*
(Burung Sewah Tahu) 42cm

DESCRIPTION Large bird with a red iris. Males generally glossy black. Females mainly dark brown with spots on head, upperparts and barred upper tail; light brown underparts with dark brown streaks and bars. **DISTRIBUTION** Found across India, China and Southeast Asia to Australia. In Malaysia and Singapore, residents are boosted by passage migrants and winter visitors. **HABITS AND**

HABITAT Seen in coastal areas, plantations, and wooded gardens in towns and cities. Generally shy and confined to the security of dense foliage, but loud calls, *Ko-el, Ko-el*, betray its presence. More vocal during the breeding period, much to the annoyance of city-dwellers. Known brood parasite of the House Crow (p. 94) in Malaysia.

FAR LEFT: *Male*. LEFT: *Female*

Bock's Hawk-cuckoo

■ *Hierococcyx bocki* (Burung Sewah Tekukur Besar) 33cm

DESCRIPTION Medium-sized cuckoo with grey-brown head and back, wings and banded tail. Underparts white, with rich orange breast, variably streaked; the white flanks with well-spaced bars. Yellow eye-ring and legs. **DISTRIBUTION** Resident in mountains of Peninsular Malaysia, Sumatra and Borneo. This species recently split off from **Large Hawk-cuckoo** *H. sparverioides* from Himalayas to S China (bigger, duller, with black chin and streaked throat), an uncommon migrant to highlands and lowlands of Southeast Asia including Peninsular Malaysia, Singapore, Sabah and Sarawak. **HABITS AND HABITAT** Fairly common but secretive resident in montane forest, 900–1,800m, usually solitary, and best detected by its advertising call, a series of disyllables *pi-pi*; *pi-pi*; … that gradually rise to become frantic.

Malaysian Hawk-cuckoo

■ *Hierococcyx fugax* (Burung Sewah Hantu) 29cm

DESCRIPTION Dark grey-brown head with a little white on lores; traces of a white hind-collar; dull, dark brown upperparts; and 3 or 4 black bars (the last the widest) across ashy-grey tail. Underparts creamy white with bold black streaks. The migrant form, with grey upperparts and lighter streaking below, is now separated as the species Hodgson's Hawk-cuckoo *H. nisicolor*. **DISTRIBUTION** Resident in Peninsular Malaysia, Borneo (including Sabah and Sarawak) and Sumatra, with several records but no proven breeding in Singapore. **HABITS AND HABITAT** Found in lowland forest, both pristine and partly disturbed, on coastal plains up to *c.* 250m, keeping to the middle and lower storeys. A parasitic species, possibly laying in nests of shamas, but more details of host species are needed. Calls include a succession of 2 notes on the same pitch, with emphasis on each 1st note; and a succession of 2 notes that gradually rise and accelerate, with emphasis on each 2nd note, before breaking up into bubbling sounds.

Chestnut-bellied Malkoha
■ *Phaenicophaeus sumatranus* (Burung Cenuk Kecil) 40cm

DESCRIPTION Large, elongated bird with a light grey head, throat and chest. Prominent red orbital skin; light green bill; glossy, dark green upperparts and wings; inconspicuous

dark cinnamon belly and vent. Long tail has white tips. **DISTRIBUTION** From *c.* 12°N in Myanmar, through the Malay Peninsula to Borneo and Sumatra. Resident in Peninsular Malaysia, Sabah, Sarawak and Singapore. **HABITS AND HABITAT** Generally unobtrusive in lowland habitats such as mangroves, primary forests and forest edges, secondary forests and plantations. Forages in the middle storey of the vegetation in search of insect prey, peering around to locate large insects, caterpillars, small lizards, and some seeds or fruits. The only malkoha species now remaining in Singapore.

Green-billed Malkoha
■ *Phaenicophaeus tristis* (Burung Cenuk Kera) 55cm

DESCRIPTION Large and grey, with a very long, white-tipped tail. Similar to the Chestnut-bellied Malkoha (above), but with a paler breast and grey (not brown) belly and vent. Green bill and red skin around eye. **DISTRIBUTION** From the Himalayas and NE India through S China and S to Peninsular Malaysia, Sumatra and Kangean. Resident in the N half of Peninsular Malaysia, above *c.* 3°N. Not known from Singapore, Sabah or Sarawak. **HABITS AND HABITAT** Similar in habits to other malkohas, foraging for large insect prey in dense foliage around tree trunks. In Peninsular Malaysia it prefers montane forest above *c.* 850m, although in Thailand and further northwards it occurs in a wider range of habitats down to coastal mangroves, bamboo groves, orchards and plantations.

Red-billed Malkoha
■ *Phaenicophaeus javanicus*
(Burung Cenuk Api) 45cm

DESCRIPTION Ashy grey above and rich cinnamon-fawn from chin to vent, with grey flanks; long, dark grey tail with white tips. The only malkoha with an entirely red bill; small area of blue skin around eye. **DISTRIBUTION** From *c.* 14°N in Myanmar through the Malay Peninsula to Sumatra, Borneo and Java. Resident in Peninsular Malaysia, Sabah and Sarawak, but now locally extinct in Singapore. **HABITS AND HABITAT** Similar to other malkohas, occurring in forests and forest edges from the lowlands up to *c.* 1,200m in lower montane forest. There is little information about its diet, and the means of ecological separation between different malkohas would be a useful study topic.

Chestnut-breasted Malkoha
■ *Phaenicophaeus curvirostris* (Burung Cenuk Birah) 45cm

DESCRIPTION Large malkoha with greyish head, and rufous-brown throat and underparts all the way to vent; glossy, dark green upperparts, wings and slightly more than half of long tail (distal part is rufous). Prominent

red eye-patch, the red colour continuing across lower mandible; yellow iris. **DISTRIBUTION** From *c.* 15°N in Myanmar and Thailand through the Malay Peninsula to the Greater Sundas. Resident in Peninsular Malaysia, Sabah and Sarawak, but now locally extinct in Singapore. **HABITS AND HABITAT** Prefers forests, forest edges, plantations and wooded gardens. Usually in pairs and generally unobtrusive.

Greater Coucal

■ *Centropus sinensis*
(Burung Butbut Cari
Anak) 52cm

DESCRIPTION Large,
crow-like bird with heavy,
clumsy flight. Plumage
glossy black overall, wings
chestnut, eyes red.
DISTRIBUTION From
India through S China to
Southeast Asia. Resident
in Peninsular Malaysia,
Sabah, Sarawak and
Singapore. **HABITS
AND HABITAT** Often
encountered at forest
edges, scrub, riverine
vegetation and plantations,
singly or, rarely, in pairs.
Very shy and confined to
thick vegetation, showing
itself while sunning or
scrambling among the
foliage. Call is a prolonged
series of deep booms.

Lesser Coucal

■ *Centropus bengalensis* (Burung Butbut Kecil) 37cm

DESCRIPTION Generally black with chestnut
wings; similar to Greater Coucal (above), but smaller
and with a variable amount of pale streaks on head,
throat, chest and wings (Greater Coucal never
streaked). **DISTRIBUTION** From India through S
China to Southeast Asia. Resident in Peninsular
Malaysia, Singapore, Sabah and Sarawak. **HABITS
AND HABITAT** Common resident in the lowlands
up to *c.* 1,500m. Often encountered in grasslands and
scrub, foraging in dense shrubs or on the ground, or
sunning with wings drooping or outstretched. Makes
occasional low, short flights over vegetation. Typical
call (one of several) is a guttural *ko-kok, ko-kok....*

Barn Owl ■ *Tyto alba* (Burung Pungguk Jelapang) 35cm

DESCRIPTION One of the most easily recognisable owls, with a white heart-shaped face and buffy rim. Throat and underparts white; head, upperparts and tail buffy with small whitish spots; legs long. **DISTRIBUTION** Virtually global distribution. Common resident in Peninsular Malaysia and Singapore since the 1960s, but recorded in Sabah and Sarawak only as a result of deliberate human introductions. **HABITS AND HABITAT** Usually seen solitary in low flight, or perched in open country, rice fields, oil-palm plantations and cultivation. Used as a biological agent in rice fields and plantations to control rodent populations. Roosts in trees and uninhabited buildings, or in nestboxes provided by landowners to boost numbers of the species.

Collared Scops-owl ■ *Otus lempiji* (Burung Hantu Reban) 21cm

DESCRIPTION Small rufous-brown owl with short ear tufts; back speckled, mottled and spotted darker; underparts faintly vermiculated, each feather with a dark spot near tip. Fronts of ear tufts and a variable broad collar around neck cream to deep warm buff. Eyes dark brown. **DISTRIBUTION** Resident from India and Sri Lanka to Siberia and Japan, S through Southeast Asia to Java and Bali. Resident in Peninsular Malaysia, Singapore, Sabah and Sarawak. **HABITS AND HABITAT** Keeps to tall secondary woodland, tree plantations, well-wooded gardens and, less commonly, undisturbed forest, hunting for beetles, cockroaches, crickets, small lizards and other small vertebrates at night. Call is a soft *po-up*, on 1 pitch or deflected downwards on the 2nd note, and repeated at regular intervals of c. 12 seconds; from a distance, it sounds like a single note.

Barred Eagle-owl ■ *Bubo sumatranus* (Burung Hantu Bubu) 45cm

DESCRIPTION Large bird, but not reaching size of European and American eagle-owls. Mottled and barred buff on dark brown upperparts, and broadly, closely barred below on white underparts. Big, dark eyes; long ear tufts give it a rather flat-crowned appearance. Juvenile plumage contains a lot of white.
DISTRIBUTION From *c.* 13°N in Thailand, through the Malay Peninsula to Sumatra, Borneo, Java and Bali. Resident in Peninsular Malaysia, Sabah and Sarawak; former resident, now locally extinct, in Singapore. **HABITS AND HABITAT** Typically solitary, in the middle and upper storeys of undisturbed forest in lowlands, from plains level to *c.* 900m, rarely higher. Typical call is a deep, 2-note *huh huh*, repeated at intervals.

Buffy Fish-owl
■ *Bubo ketupu* (Burung Hantu Kuning) 45cm

DESCRIPTION Large, bright buffy-cinnamon owl; dark streaks on underparts; darker brown above with buff streaks and bars. Conspicuous angular ear tufts; pale patch above bill, and staring yellow eyes.
DISTRIBUTION From NW India and Myanmar through Indochina to Laos and Vietnam, southwards through the Malay Peninsula to Sumatra, Borneo, Java and Bali. Resident in Peninsular Malaysia, Singapore, Sabah and Sarawak. **HABITS AND HABITAT** Usually forest edges close to rivers or lakes, including both natural and artificial ponds; perches on a low branch at water's edge, waiting for prey, such as fish, frogs and other small animals. Also found in lowlands in mangroves, orchards and plantations. Calls include harsh, hair-raising wails.

Spotted Wood-owl
■ *Strix seloputo* (Burung Hantu Carik Kafan) 46cm

DESCRIPTION Distinctive orange-brown
facial disc, topped by dark brown crown with
whitish spots; dark brown upperparts with white
spots, and pale underparts with regular bars.
Eyes contrasting dark brown against facial disc.
DISTRIBUTION SE of a line from N Thailand
to southernmost Vietnam, through the Malay
Peninsula to Sumatra, Java and the Philippines,
including Palawan, but absent from Borneo.
Resident in Peninsular Malaysia and, since 1985,
in Singapore. **HABITS AND HABITAT** Forest
edges, plantations, tall secondary woodland and
thickly tree-dominated gardens, avoiding the
interior of undisturbed forest. Rats and other
small vertebrates are the usual food. Typical
call is a single powerful note, *huh*, repeated at
intervals of 10–20 seconds, at night and often
just before dawn.

Bornean Wood Owl ■ *Strix leptogrammica* (Burung Hantu Punggur) 45cm

DESCRIPTION Similar to the Barred Eagle-owl (p. 54), but with a rounded head
lacking ear tufts, and a distinct
rufous mask outlined with a dark
surround and darkening to smudges
around dark eyes. Finely barred
below, and lacking spots on crown.
DISTRIBUTION Borneo (including
Sabah, Sarawak) and Java. Very
similar **Brown Wood Owl** *Strix
indranee* occurs from South Asia to
Peninsular Malaysia and Sumatra.
HABITS AND HABITAT Occurs
in forest interiors, from the extreme
lowlands up through lower montane
forest to *c*. 1,700m. Will sometimes
come to the forest edge, but usually
waits on a branch in the middle
storey, looking and listening for prey.
Call, around dusk and at night, is a
wavering, deep *huhuhooo*.

Brown Boobook

■ *Ninox scutulata* (Burung Hantu Betemak) 30cm

DESCRIPTION Larger than a scops-owl, with a rounded head, no ear tufts and round, staring yellow eyes. Plumage dark brown, including face, with some pale spotting on upperparts, and the brown increasingly broken up by white on lower breast and belly; tail barred. **DISTRIBUTION** From India and Sri Lanka across E Asia to Korea and Japan, and through Southeast Asia to Sumatra, Borneo, Java, Bali and the Philippines. Of these, N continental Asian birds are now split as the separate species Northern Boobook *N. japonica*. Resident in Peninsular Malaysia, Singapore, Sabah and Sarawak, and also occurs as a migrant throughout the region. **HABITS AND HABITAT** Resident in lowland forest, forest edges and tall secondary woodland at low altitudes. Hunts primarily for insects, and can sometimes be seen by day. Most often detected by frequent calling, *ke-wick*. The migrant form occurs at a wider range of altitudes up to 2,000m.

Gould's Frogmouth ■ *Batrachostomus stellatus* (Burung Segan Bintik Mas) 22cm

DESCRIPTION Immensely wide gape, big eyes and mottled brown plumage are typical of frogmouths. Underparts buffy white, with darker edges forming regular scallops; several lines of small, black-edged white spots on wing coverts and a larger series close to scapulars; large yellow eyes. Dark brown, rufous-brown and, rarely, greyish-brown phases occur. **DISTRIBUTION** From *c.* 10°N in the Malay Peninsula through Sumatra, Borneo and the larger intervening islands. Resident in Peninsular Malaysia, Sabah and Sarawak; former resident in Singapore, now locally extinct. **HABITS AND HABITAT** Within tall forest in the extreme lowlands, up to 200m. Roosts on a branch in the middle storey by day, well camouflaged, and flies at night to catch insects on the wing. Male's call is a soft, 3-part *ooo-tuiloo-kuk*, with the last part either rising or falling.

Large-tailed Nightjar ■ *Caprimulgus macrurus* (Burung Tukang Kubur) 32cm

DESCRIPTION Beautifully camouflaged, mainly brown bird, mottled with black and grey, and with wide greyish sides to crown. Male has white throat, white patches on primaries and white corners to tail, all of which are present but more buff-coloured in female. **DISTRIBUTION** From NE India through S China to Hainan, and southwards through Southeast Asia to the Philippines, Greater Sundas and N Australia; there is some

disagreement in defining species limits through part of this range. Resident in Peninsular Malaysia, Singapore, Sabah and Sarawak. **HABITS AND HABITAT** Very common in lowland habitats, in areas with tall secondary woodland, gardens, and open ground with mixed scrub. Perches on fences, lamp-posts, etc., or even on the ground, and hawks for prey over open stretches. Often detected by its call, a repetitive *klok, klok, klok…*, like someone knocking on wood, continued in bursts for long periods of the night.

Savanna Nightjar ■ *Caprimulgus affinis* (Burung Tukang Padang) 25cm

DESCRIPTION Small nightjar, with a shorter tail and finer mottling than Large-tailed Nightjar (above), and without bold black markings. Male has white wing-patches and mostly white outer tail feathers. Female has buff wing-patches, and outer tail feathers do not differ from rest of tail. **DISTRIBUTION** From the Himalayas through southernmost

China to Taiwan, and Southeast Asia to Sumatra, Java, Bali, parts of Borneo, Sulawesi and the Lesser Sundas. Resident and spreading in Peninsular Malaysia and Singapore since the 1980s, now extending in Borneo and perhaps reaching Sabah. **HABITS AND HABITAT** Typical of more open grassland than the Large-tailed, including grassy dunes near coasts and reclaimed land. Call is an explosive *chewee*, given in flight, usually for a fairly short period around and just after dusk.

Plume-toed Swiftlet ■ *Collocalia affinis* (Burung Layang-layang Licin) 10cm

DESCRIPTION Smaller than House Swift (p. 59), and glossy blue-black overall; greyish chin and dirty whitish belly without clearly defined margins; tail slightly notched.

Separated from other swiftlets in flight by its small size and habit of flying very close to the surface of vegetation such as tree crowns. **DISTRIBUTION** Peninsular Malaysia to Singapore, Sumatra, Borneo and offshore islands. Resident from extreme lowlands to highlands in Peninsular Malaysia, Sabah and Sarawak; formerly resident in Singapore but now apparently only a visitor. **HABITS AND HABITAT** Often seen in flight in small flocks over forests, forest edges, open country, towns and cities. Occasionally, will skim over rivers or pools to drink. Nests in cave mouths, under eaves, and in tunnels and similar structures where light penetrates.

Himalayan Swiftlet ■ *Aerodramus brevirostris* (Burung Layang-layang Himalaya) 14cm

DESCRIPTION Field identification of swiftlets is extremely difficult. Himalayan has a moderate tail notch and pale rump, and is said to have a stiff-winged, flicking flight. In comparison, **Edible-nest Swiftlet** *A. fuciphagus* is smaller with a moderate tail notch and

pale rump; and **Black-nest Swiftlet** *A. maximus* is slightly larger and heavier-looking with broader wings and less of a tail notch. **DISTRIBUTION** Himalayan is resident from the Himalayas through to Hubei in central China, and southwards to N Thailand; also in Java. Passage migrant and non-breeding visitor in Peninsular Malaysia, and likely in Singapore (with some disagreement over identification criteria; not known from Sabah or Sarawak. Edible-nest is resident from the Andamans and S Myanmar through Peninsular Malaysia, Sumatra, Borneo and Java to parts of the Lesser Sundas. Black-nest is resident from S Myanmar through Peninsular Malaysia to Sumatra, Borneo and Java. **HABITS AND HABITAT** All 3 of these swiftlet species spend most of their time flying, from low to very high in the sky and at all altitudes, from the extreme lowlands to mountain tops, over forest, and in disturbed and open habitats.

House Swift ■ *Apus nipalensis*
(Burung Layang-layang Rumah) 15cm

DESCRIPTION Generally medium-sized bird with glossy black plumage and prominent white rump; throat white; tail slightly notched. **DISTRIBUTION** Africa, Middle East, India and S China to Southeast Asia and the Greater Sundas. Resident in Peninsular Malaysia, Singapore, Sabah and Sarawak. **HABITS AND HABITAT** Prefers open country, forest edges, towns and cities. Gregarious by nature, often forming large breeding colonies under eaves of man-made structures such as buildings or bridges, and on cliffs or cave mouths. Harsh, trilling screams are given around roosting and nesting sites.

Whiskered Treeswift ■ *Hemiprocne comata*
(Layang-layang Berjambul Kecil) 16cm

DESCRIPTION Small, slim brown bird, with long wings and deeply forked tail. Dark brown body with 2 white lines on side of head (brow and moustache), and white on innermost wing feathers. Ear coverts between the 2 white lines are maroon in male, blackish in female. **DISTRIBUTION** From c. 12°N in southernmost Myanmar through the Malay Peninsula to Sumatra, Borneo and the Philippines. Resident in Peninsular Malaysia, Sabah and Sarawak; considered formerly resident in Singapore but now only a rare visitor. **HABITS AND HABITAT** Often seen perched on the end twigs of tree crowns, with wing-tips crossed over rump, or in flight over lowland forest and just into montane forest to c. 1,100m. Makes short flights after insects, typically returning to the same perch.

LEFT: *Female*. RIGHT: *Male*

Red-naped Trogon ■ *Harpactes kasumba* (Burung Kesumba Batang) 32cm

DESCRIPTION Male has a narrow white line separating scarlet breast from black upper breast and throat, and a broad scarlet patch behind neck, this meeting the blue facial skin

that is contiguous with the blue bill. Female has a brownish-buff lower breast, sharply defined from dark brown upper breast. In both sexes, tail feathers are tipped by unmarked white. **DISTRIBUTION** From *c.* 8°N in Peninsular Thailand through to Borneo and Sumatra. Resident in Peninsular Malaysia, Sabah and Sarawak; former resident in Singapore, now locally extinct. **HABITS AND HABITAT** In the middle and upper storeys of tall lowland forest, to *c.* 600m on hill slopes, occasionally higher. Insects are snatched from the foliage. Call is a mellow *taup taup taup taup*, usually only 3 or 4 notes; generally lower in pitch and delivered more slowly by male than female.

Diard's Trogon ■ *Harpactes diardii* (Burung Kesumba Diard) 33cm

DESCRIPTION Male has a pink band separating scarlet breast from black upper breast and throat; and a broad pink patch behind neck, not meeting the violet facial skin that is itself separated from the blue bill. Female has a uniform brown head and upper breast,

distinct from reddish-pink lower breast and belly. In both sexes, white tips of tail feathers are vermiculated with black. **DISTRIBUTION** From *c.* 9°N in Peninsular Thailand through to Borneo and Sumatra. Resident in Peninsular Malaysia, Sabah and Sarawak; former resident in Singapore, now locally extinct. **HABITS AND HABITAT** In the middle and lower storeys of tall lowland forest, to *c.* 900m on hill slopes. Insects are snatched from the foliage. Call is a resonant series of 10–12 notes, delivered fast, on a falling pitch after the 2nd note, and accelerating.

Scarlet-rumped Trogon

■ *Harpactes duvaucelii* (Burung Kesumba Puteri) 25cm

DESCRIPTION Smaller trogon. Male has a black throat, bright scarlet upper and lower breast, and extensive bright scarlet rump. Female has brown head and throat, buff breast merging into pinkish belly, and rufous rump with traces of pink. In both sexes, blue skin on head is limited to a projecting line on brow, and at base of bill is a bright blue gape. **DISTRIBUTION** From *c*. 14°N in Myanmar through the Malay Peninsula to Borneo and Sumatra. Resident in Peninsular Malaysia, Sabah and Sarawak; never reliably recorded from Singapore. **HABITS AND HABITAT** In the middle and lower storeys of tall lowland forest, to above 1,000m on hill slopes. Insects are snatched from the foliage and, occasionally, in flight. Call is a rapid series of 10–12 notes, delivered so fast that they run together as they accelerate.

Orange-breasted Trogon

■ *Harpactes oreskios* (Burung Kesumba Harimau) 29cm

DESCRIPTION Medium-small trogon. Olive head and upper breast, greener in male and browner in female; brown upperparts, more chestnut in male; orange-yellow lower breast. Around eye is a small circle of blue skin. **DISTRIBUTION** From SW China through Myanmar, mainland Southeast Asia and the Malay Peninsula to Sumatra, Borneo and Java. Resident in Peninsular Malaysia, Sabah and Sarawak; not recorded from Singapore. **HABITS AND HABITAT** In the middle and lower storeys of tall lowland and lower montane forest, from sea-level to *c*. 1,300m on mountain slopes. Like other trogons, it builds its nest in rotten stumps. Call is introduced by 1 or 2 slow, separate notes, followed by a series of 3 or 4 quick notes on 1 pitch. Insects are snatched from the foliage.

Red-headed Trogon ▪ *Harpactes erythrocephalus* (Burung Kesumba Gunung) 33cm

DESCRIPTION Large trogon with diagnostic red head in males and cinnamon head in females. Both sexes have reddish underparts with a white 'crescent' on chest (sometimes hidden between feathers), cinnamon back and upper tail, and black wings with whitish stripes on wing coverts. **DISTRIBUTION** Resident from the Himalaya to S China, Southeast Asia and Sumatra. Resident in the Main Range, Larut Range and Mount Benom in Peninsular Malaysia; not in Singapore, Sabah and Sarawak. **HABITS AND HABITAT** Prefers hill forests at 700m, and usually seen in the middle storey. Generally unobtrusive. Has been recorded participating in mixed-species feeding flocks.

Whitehead's Trogon

▪ *Harpactes whiteheadi* (Burung Kesumba Kinabalu) 33cm

DESCRIPTION Male has a brilliant scarlet crown, dark blue facial skin and black throat, scarlet lower breast and belly. Female has cinnamon head and belly. In both sexes, black throat shades gradually into light grey and almost white on upper breast; cinnamon upperparts and black wings; wing coverts with narrow whitish stripes. **DISTRIBUTION** Endemic to N Borneo, in high mountains from Mount Kinabalu in Sabah westwards to Mount Mulu, Mount Dulit and Usun Apau in Sarawak. **HABITS AND HABITAT** Montane forest, usually above 1,000m. Often seen perched in the lower storey, from where it sallies out to snatch insects from nearby foliage. Generally silent.

Rufous-collared Kingfisher

■ *Actenoides concretus* (Burung Pekaka Rimba) 24cm

DESCRIPTION Forest kingfisher with greenish crown, bold black stripe through eye, blue-black moustache, and entirely rufous-buff underparts. Male is brighter overall, with glossy blue back and wings; in female, upperparts are dull green with buff speckles on wings. **DISTRIBUTION** From *c.* 11°30'N in Myanmar and Thailand, through the Malay Peninsula to Borneo and Sumatra. Resident in Peninsular Malaysia, Sabah and Sarawak; formerly resident in Singapore, now locally extinct. **HABITS AND HABITAT** Found in the middle and, especially, lower storeys of lowland forest from sea-level up into lower montane forest at *c.* 1,200m. Perches motionless until it spots an insect or small lizard or snake. Usual call is a wavering upward whistle, delivered in a long series.

Stork-billed Kingfisher ■ *Pelargopsis capensis* (Burung Pekaka Buaya) 37cm

DESCRIPTION Large kingfisher with a massive red bill. Brown head with light orange/rufous nape, throat, underparts and vent; dull bluish-green wings and tail; red feet.

DISTRIBUTION Resident from India to Southeast Asia, usually at low elevations. **HABITS AND HABITAT** Solitary and rather silent kingfisher, often spotted perched in mangroves, former tin-mining wetlands and forested river courses. Diet consists mainly of fish, but will not hesitate to take insects and amphibians.

White-throated Kingfisher ■
Halcyon smyrnensis (Burung Pekaka Dusun) 28cm

DESCRIPTION Brown head and belly; white throat and breast, resembling a bib; iridescent blue back and tail; red bill and feet; black upperwing coverts. **DISTRIBUTION** Widespread from Middle East through India to China and Southeast Asia. The most common resident kingfisher in Peninsular Malaysia and Singapore. **HABITS AND HABITAT** Found in a wide range of habitats near human habitation, such as mangroves, agricultural areas, plantations, gardens and urban areas. Diet is varied, ranging from insects to amphibians. Often solitary on exposed perches, its calls announcing its presence, a loud but mellow trill, *kikikiki…*.

Black-capped Kingfisher ■ *Halcyon pileata*
(Burung Pekaka Kopiah Hitam) 30cm

DESCRIPTION Medium-sized kingfisher with a black head and upperwing coverts; bright bluish mantle, upperparts and upper tail. White collar, throat (sometimes with slight scaling) and breast grading into a rufous belly and vent; bright red bill and feet. **DISTRIBUTION** Found mainly in India, Myanmar, China and Korea. Populations in Southeast Asia, including Malaysia, are either passage migrants or winter visitors. **HABITS AND HABITAT** Usually seen individually at coastal wetlands such as mangroves, estuaries and large rivers in lowlands up to 1,200m. Not as vocal as Collared Kingfisher (p. 65).

Collared Kingfisher ■ *Todiramphus chloris* (Burung Pekaka Bakau) 25cm

DESCRIPTION Unmistakable bird with bluish-green head, upperparts and wings, and a white collar adjoining white underparts. Black eye-stripe not prominent at times. Bill grey and flesh-coloured. Might be confused with the Sacred Kingfisher *T. sanctus*, which occurs in the same habitats but is quite rare. **DISTRIBUTION** Recorded from S and Southeast Asia through New Guinea and Australia. Resident populations in Malaysia are augmented by seasonal migrants. **HABITS AND HABITAT** Usually encountered in mangroves, beach scrub and plantations. Has a varied diet, ranging from small crustaceans to reptiles and amphibians inland. Probably the most vocal of all kingfishers in Malaysia, with harsh, loud territorial calls typically consisting of repeated clusters of 2–4 shrieks.

Rufous-backed Kingfisher ■ *Ceyx rufidorsa* (Burung Pekaka Sepah) 13cm

DESCRIPTION Tiny forest kingfisher with brilliant orange-red bill and feet; rufous head, back and wings; yellow breast; white patch behind ear coverts. Belongs to a complex whose members range to the Black-backed Kingfisher *C. erithaca*, with a black mantle, deep blue wing coverts, and deep blue patches on the forehead and ear coverts; many intermediates occur. **DISTRIBUTION** Rufous-backed birds occur from the Malay Peninsula to Sumatra, Borneo, Java, the Philippines and Lesser Sundas. Black-backed birds range from India and Indochina as far S as Java. Rufous-backed birds are resident in Peninsular Malaysia, with visitors to Singapore; Black-backed birds are migrants. In Sabah and Sarawak, all resident populations are much more mixed. **HABITS AND HABITAT** In the lower storey of lowland forest, forest edges and mangroves, feeding at and near forest streams on insects and worms.

Blue-eared Kingfisher ▪ *Alcedo meninting* (Burung Pekaka Bintik-bintik) 16cm

DESCRIPTION Small kingfisher with deep iridescent blue head and wings, metallic light blue on back and tail, white ear-patch, white chin, rufous underparts, red legs; bill may be

red or black with red near base. **DISTRIBUTION** Found across India and Southeast Asia. Uncommon resident in Malaysia and Singapore. **HABITS AND HABITAT** Frequents forested streams, rivers, lakes and mangroves. Often seen perched low down, overlooking water to hunt small fishes, which it secures with lightning dives before returning to the same perch. Generally shy but can be seen flying low over the water's surface.

Common Kingfisher ▪ *Alcedo atthis* (Burung Pekaka Citcit) 18cm

DESCRIPTION One of the smallest kingfishers in the region. Head, mantle and wings bluish; back iridescent blue; ear coverts and underparts rufous; white behind ear coverts

and chin; bill black with tinges of orange; legs red. **DISTRIBUTION** Found across N Africa, Europe and temperate Asia to Southeast Asia. Resident and winter visitor to Peninsular Malaysia and Singapore. In Borneo, it is a rare winter visitor. **HABITS AND HABITAT** Owing to its size and habits, often hard to detect in its preferred habitats of forested streams, mangroves and former tin-mining wetlands. Occasionally seen flitting across the water, giving a high-pitched *peep*, or perched silently low down overlooking water in pursuit of prey.

Common Dollarbird ■ *Eurystomus orientalis* (Tiong Batu) 30cm

DESCRIPTION Dark brown head with prominent red bill and feet; dark bluish-green body

with some bluish-purple streaks at neck. In flight, shows white patches on wings. **DISTRIBUTION** Resident from India to Southeast Asia and Australia; populations are augmented by migrants from the N. **HABITS AND HABITAT** Solitary birds are commonly encountered in mangroves, beach scrub, plantations and open country, usually on prominent vantage points. Often sallies forth from these perches to hunt for prey of winged insects such as ants and termites.

Red-bearded Bee-eater ■ *Nyctyornis amictus* (Burung Berek-berek Janggut Merah) 33cm

DESCRIPTION Generally, plumpish green bird with red throat and breast resembling a 'beard'; lilac crown, orangey-red iris and grey feet. Bill thin, almost sickle-like. **DISTRIBUTION** Resident from Myanmar to Sumatra and Borneo, from the lowlands up to approximately 1,100m. **HABITS AND HABITAT** Typically found in forests or forest edges on a perch, from which it sallies forth in pursuit of insects such as bees, wasps, termites and butterflies. Usually solitary. Excavates burrows in the banks of streams to nest.

Blue-tailed Bee-eater

▪ *Merops philippinus* (Burung Berek-berek Sawah) 24cm plus tail spikes

DESCRIPTION Generally light green on upperparts, wings and underparts (grading into light blue at vent); white patch below bill; throat brownish yellow; black eye-stripe with red iris; conspicuous bright blue tail with central streamer. **DISTRIBUTION** Found from India and S China through Southeast Asia and New Guinea. Populations in Malaysia and Singapore are mainly passage migrants and winter visitors. **HABITS AND HABITAT** Usually seen in open areas, including beach scrub, former tin-mining areas and rice fields. Individuals or groups often seen perched on utility lines, using these to sally forth in pursuit of winged insects. Also uses the same perch to disarm prey such as bees of their sting. Roosts in groups.

Blue-throated Bee-eater

▪ *Merops viridis* (Burung Berek-berek Pirus) 23cm plus tail spikes

DESCRIPTION Similar in size to Blue-tailed Bee-eater (above). Brown cap, nape and mantle; black eye-stripe with red iris; bluish throat, blending into light green chest and underparts; darker green wing coverts, bluish-green primaries and tail. **DISTRIBUTION** Found from S China to Southeast Asia. Common breeding visitor in Malaysia. **HABITS AND HABITAT** Frequents open habitats, such as beach scrub, former tin-mining areas and rice fields, perched on utility lines. Forms communal roosts and nests underground, usually in areas with little grassy vegetation.

Oriental Pied Hornbill

■ *Anthracoceros albirostris* (Burung Enggang Kelingking) 68–70cm

DESCRIPTION This small hornbill is one of the commoner species. Black head, neck and wings; white lower breast and outer tail feathers. Bare, pale skin patches around eye and gape. Bill and casque are ivory-coloured, with obscure dark patches in female, and fewer but more defined, intense black marks in male. **DISTRIBUTION** Resident from India to SW China and throughout Southeast Asia to Sumatra, Borneo and Java. **HABITS AND HABITAT** In lowland forest edges, mangroves and along rivers, seldom above 500m. Like other hornbills, it nests in tree cavities, the female being sealed in with mud except for a narrow slit through which the male passes food to mother and offspring. Exciting recent work with cameras and satellite tracking has revealed many details of the species' breeding cycle, including its astonishing range of animal prey.

Black Hornbill ■ *Anthracoceros malayanus* (Burung Enggang Gatalbirah) 75cm

DESCRIPTION Entirely black except for white outer corners to tail and, in some individuals, a grey or white eyebrow stripe. Male has ivory bill and black facial skin; female has black bill and dull pink skin around eye. **DISTRIBUTION** Resident in Borneo (Sabah, Sarawak, Brunei, Kalimantan), Sumatra, Bangka, Belitung, Singkep and Peninsular Malaysia northwards to *c.* 8°N in Thailand. **HABITS AND HABITAT** In the middle and upper storeys of lowland evergreen rainforest, typically in extreme lowlands over level ground, rarely to 600m. Territorial and accompanied by previous young, searching for many kinds of fruits and, occasionally, small animals. Breeds monogamously, usually without helpers (see Bushy-crested Hornbill, p. 71). Pairs give a harsh rasping or vomiting call.

Female

Great Hornbill ■ *Buceros bicornis* (Burung Enggang Papan) 110–120cm

DESCRIPTION One of the bulkiest hornbills, easily recognised by its pied appearance, with head and white parts of plumage often stained yellow by oils from preen gland.

Male and female are alike, except for eye colour (red in males, white in females). Casque tends to be larger in males, with black trimmings. **DISTRIBUTION** Resident in parts of India, Bangladesh and Myanmar to SW China and southwards through Peninsular Malaysia to *c.* 3°N. Absent from the S part of the Malay Peninsula and Singapore, and from Borneo, but reappears in Sumatra. **HABITS AND HABITAT** In lowland and hill forest from sea-level to *c.* 1,300m. Typically occurs in pairs, but sometimes gathers in larger groups at good food sources, such as heavily fruiting fig trees. Call is a loud barking, the male and female alternating, either when perched or in flight.

Rhinoceros Hornbill ■ *Buceros rhinoceros* (Burung Enggang Badak) 90–120cm

DESCRIPTION Enormous black hornbill with white belly, white tail crossed by a black bar, and brilliant yellow and red bill and casque. Male has larger casque with black line, and a red eye; female has white eye surrounded by red skin. **DISTRIBUTION** Resident in suitable habitats throughout Borneo, Java, Sumatra and Peninsular Malaysia, to *c.* 7°N in Thailand; historically resident in Singapore. **HABITS AND HABITAT** Canopy of lowland evergreen rainforest, to 1,400m. Typically occurs in pairs and, sometimes, with previous young. Flocks of up to 25 occur rarely at good fruiting trees. Monogamous pair breeds without helpers (see Bushy-crested Hornbill, p. 71), in natural tree cavities. Members of the pair advertise their territory and keep in touch with a loud, nasal, barking duet, *eng – gang*, often in flight.

Bushy-crested Hornbill ■ *Anorrhinus galeritus* (Burung Enggang Mengilai) 90cm

DESCRIPTION Dark grey-brown all over; darkest on head, becoming paler down to wings and basal half of tail, with distal half of tail forming blackish band. Male has blue face and black bill; female has pink face and particoloured bill. **DISTRIBUTION** Resident in Borneo (Sabah, Sarawak, Brunei, Kalimantan), Natuna Besar, Sumatra and the Malay Peninsula, northwards through Peninsular Thailand and Myanmar to *c.* 14°N. **HABITS AND HABITAT** In lowland and, occasionally, lower montane forest, from sea-level to *c.* 1,400m or more, but commonest in foothills. Noisy groups of adults and their young, giving raucous, puppy-like yelping choruses, seek a variety of fruits, including some figs, and small invertebrates. Nesting takes place in a tree cavity, helped by group members – often a previous batch of young, which assist by bringing food to the nest.

White-crowned Hornbill
■ *Berenicornis comatus* (Burung Enggang Jambul Putih) 90–100cm

DESCRIPTION Perhaps the hornbill that is least often seen. Distinctive shaggy white head (crown only in female) and tail, and entirely black wings except for white trailing edge; males also have a white breast. Juveniles dark all over with white-speckled head, and only distal half of tail is white. **DISTRIBUTION** Resident patchily throughout Borneo (Sabah, Sarawak, Brunei, Kalimantan), Sumatra and the Malay Peninsula, northwards in Peninsular Thailand to *c.* 15°N and in Myanmar to 14°N. **HABITS AND HABITAT** In lowland evergreen rainforest to 900m. Typically in the lower to middle storeys in small territorial groups usually including an adult pair with helpers (see Bushy-crested Hornbill, above) and juveniles. Feeds on lizards, snakes, small birds, bats and large insects, as well as fruits but rather few figs. Call is a soft ventriloqual cooing.

Fire-tufted Barbet ■ *Psilopogon pyrolophus* (Burung Takur Jambang Api) 28cm

DESCRIPTION Large barbet, grass-green on body and wings; dark collar, above which

throat is yellow and ear coverts are greyish white; crown dark, with fiery chestnut tuft above pale greenish bill, this crossed by a black band. Sexes are alike. **DISTRIBUTION** Resident in mountains of Sumatra and Peninsular Malaysia. **HABITS AND HABITAT** Found in the canopy and middle storey of montane evergreen rainforest at *c.* 900–2,000m, and sometimes down into the understorey along disturbed forest edges. Feeds on many kinds of figs and other soft fruits, plus a few insects. Often located by its weird rasping call, the notes accelerating to a whirr like a buzzing insect.

Lineated Barbet ■ *Psilopogon lineatus* (Burung Takur Kukup) 28cm

DESCRIPTION Has the typical barbet pattern of green back, wings and tail, but head and forequarters are distinctively oatmeal with brown streaking. Bill and skin around eye bright yellow. Sexes are alike. **DISTRIBUTION** Resident from the basal E Himalayas through Myanmar, Thailand and Indochina to N Peninsular Malaysia; also in Java and Bali. Now spreading southwards to fill in parts of coastal Peninsular Malaysia, and introduced to Singapore. **HABITS AND HABITAT** Open and disturbed forest and forest edges, especially near the coast. Like other barbets, it excavates holes in small trees or branches to nest; this species usually chooses living trees or may reuse old cavities. Eats mainly fruits, but fewer figs than most barbets. Call is a distinctive *ku-kruk*, repeated about once a second in a long series.

Golden-naped Barbet
▪ *Psilopogon pulcherrimus*
(Burung Takur Bintarang) 21cm

DESCRIPTION Green with light blue forehead, crown and throat, black lores, and blur of golden yellow on hind-neck in adults (lacking in juveniles). Sexes are alike. **DISTRIBUTION** Endemic to Borneo, where it is resident in mountains of Sabah and Sarawak as well as adjacent East Kalimantan. **HABITS AND HABITAT** Montane forest at *c.* 1,100–2,500m, where it seeks fruits such as figs and may also take some insects; it is a fig specialist but takes other fruits such as *Medinilla* from the middle and lower storeys. Nest is a cavity excavated in mouthfuls of rotten wood from a dead tree, often a very slim one at the forest edge. Most commonly recorded from its call, *tuk tuk tukrrrk*, given regularly every few seconds from a perch among dense foliage in smaller trees and the middle canopy. Like most other barbets, also has a serial trilling call, each trill shorter than the last.

Golden-throated Barbet ▪ *Psilopogon franklinii* (Burung Takur Gunung) 22cm

DESCRIPTION Green with yellow upper throat, greyish lower throat, red forehead and yellow hind-crown. Red spot on nape; sides of head mostly grey. **DISTRIBUTION** From Nepal and NE India to S China, and patchily southwards in mountains to Peninsular Malaysia; resident there, but not recorded in Singapore, Sabah or Sarawak. **HABITS AND HABITAT** A barbet of montane forest on the Malay Peninsula, in the Larut Range, the S part of the Main Range and scattered mountains further E. Occurs in the upper montane forest above 1,400m, feeding on figs and other fruits, although its diet is not well documented. Call is a rapid, repeated *ke-triuk, ke-triuk….*

Yellow-crowned Barbet

■ *Psilopogon henricii* (Burung Takur Mahkota Kuning) 22cm

DESCRIPTION Green with a yellow forecrown and eyebrow, and blue hind-crown and throat; black lores and narrow eye-ring. As in most other barbets, sexes are alike. **DISTRIBUTION** From *c.* 8°30'N in the Malay Peninsula, through Borneo and most of Sumatra. Resident in Peninsular Malaysia, Sabah and Sarawak, but never reliably recorded from Singapore. **HABITS AND HABITAT** The upper storey of lowland forest, from sea-level to *c.* 900m at the transition to lower montane forest, where it meets but does not overlap with Black-browed Barbet M. *oorti*. Birds gather in the canopy at fruiting figs with other barbets and frugivorous birds, but are otherwise fairly solitary. Call is a repeated *tuk tuk tuk tuk trrrk*, with 4 short notes and 1 long.

Coppersmith Barbet

■ *Psilopogon haemacephalus* (Burung Takur Tukang Besi) 17cm

DESCRIPTION The smallest barbet locally, green above and streaky buff and green below. Yellow above and below eye and on throat; red forehead and breast-band. Sexes are alike, juveniles somewhat duller. **DISTRIBUTION** Resident from the Indian sub-continent through Southeast Asia to Peninsular Malaysia, Sumatra, Java and the Philippines, but absent from the whole of Borneo. **HABITS AND HABITAT** The only barbet typical of open country with scattered trees, parkland and even urban gardens; also found in plantations and the landward side of mangroves. Single birds, or occasionally loose aggregations at fruiting figs, are often seen perched on bare, protruding canopy twigs. Calls attention to itself by a monotonous, repeated *toink, toink, toink…*, which gives the species its common name.

Sunda Pygmy Woodpecker
■ *Picoides moluccensis* (Burung Belatuk Belacan) 13cm

DESCRIPTION Small woodpecker with dark brown cap (with a red flash in males); broad, dark brown stripe through and behind eye, and another dark moustachial stripe. Upperparts dark brown, striped with white; underparts whitish with streaks. Wide, pale eyebrow is usually a striking feature in the field. **DISTRIBUTION** Resident in India and Sri Lanka, Peninsular Malaysia, Sumatra, Borneo, Java and the Lesser Sundas. **HABITS AND HABITAT** Found in mangroves, coastal woodlands, secondary growth and, in places, the trees in parks and on roadsides, always at low altitudes fairly near the coast. Sometimes comes down to the understorey, even foraging on wooden fenceposts. Usually seen singly or in pairs.

Female

Rufous Woodpecker
■ *Micropternus brachyurus* (Burung Belatuk Biji Nangka) 25cm

DESCRIPTION Overall rufous brown, slightly darker on head and tail, with black bars on back, wings and tail. No significant crest. Males have a patch of red below eye (absent in female), but this is not very conspicuous against the generally brown colour. **DISTRIBUTION** From the Himalayan foothills, E India and Sri Lanka through S China to Hainan, and southwards through Indochina to the Malay Peninsula, Sumatra, Borneo and Java. Resident in Peninsular Malaysia, Singapore, Sabah and Sarawak. **HABITS AND HABITAT** Found in lowland forest, from sea-level to c. 1,000m, where it may just enter lower montane forest; also in mangroves and tall secondary forest. Usually in tree crowns, coming lower at forest edges and in clearings.

Banded Yellownape
■ *Chrysophlegma miniaceum*
(Burung Belatuk Merah) 26cm

DESCRIPTION One of several yellow-crested woodpeckers. Rufous head and throat are diagnostic, merging into barred green back and irregularly buff- and brown-banded breast and belly. Rump yellow, wings crimson. In females, sides of face are duller with more white speckles. **DISTRIBUTION** Resident in Java, Sumatra, Nias, Borneo, Bangka, Belitung, Singapore and northwards through Peninsular Malaysia to *c.* 13°N in Peninsular Thailand. **HABITS AND HABITAT** Inhabits the middle storey of tall secondary forest, parks, gardens and forest edges, up to *c.* 1,200m. May be located by its loud, repeated scream, *kwee, kwee.*

Chequer-throated Yellownape
■ *Chrysophlegma mentale* (Burung Belatuk Ranting) 28cm

DESCRIPTION Another yellow-crested woodpecker, but set apart by greenish crown and broad zone of light chestnut running from behind eye, down neck and around

underside of throat, the throat being chequered black and white. In female, the chestnut continues as a malar stripe. **DISTRIBUTION** Resident in southernmost Myanmar and Peninsular Thailand, Peninsular Malaysia, Sumatra, Bangka, Java and Borneo; now lost from Singapore. **HABITS AND HABITAT** Found foraging on trunks and larger boughs in the middle storey of lowland and hill evergreen rainforest to *c.* 1,200 m, but scarcer than either the Crimson-winged (p. 77) or Banded (above) woodpeckers. Calls have an upward inflection, *kiyee*, compared with all those of Crimson-winged, which have a downward inflection.

Crimson-winged Woodpecker

■ *Picus puniceus* (Burung Belatuk Mas) 26cm

DESCRIPTION Similar to the Banded Woodpecker (p. 76), but sides of head and throat green (males with a red moustache streak), breast plain green with barring lower on belly, back plain green. Rump yellow and wings bright crimson. Contrast between red crown and green face is a useful feature for identification. **DISTRIBUTION** Resident in Java, Sumatra, Nias, Borneo and Peninsular Malaysia, northwards to *c.* 13°N in Peninsular Thailand. Now lost from Singapore. **HABITS AND HABITAT** More characteristic of primary lowland evergreen rainforest than the Banded Woodpecker, but also occurs in rubber and oil-palm plantations and secondary woodland. Characteristic call is a 2-note wail, *kee-bee*, the 2nd note lower; also has several other types of call, each presumably with a differing function.

Female

Common Flameback

■ *Dinopium javanense* (Burung Belatuk Pinang Muda) 30cm

DESCRIPTION Golden-brown back and wings, and bright orange-red rump (hence its common name); white underside with black scallops and bold black and white face stripes make it conspicuous. Crown and crest red in male, black in female. **DISTRIBUTION** Resident in parts of India through to S China and the whole Indo-Malayan region as far as Sumatra, Borneo, Java, Bali and parts of the Philippines. Found throughout Peninsular Malaysia, Singapore, Sabah and Sarawak. **HABITS AND HABITAT** Probes and gleans for ants, termites and other insects on the bark of big trees in lowland parkland, gardens, and timber, oil-palm and rubber plantations, and especially in mangroves. Usually seen in pairs, often calling with loud, short rattles in flight.

Green Broadbill ■ *Calyptomena viridis* (Burung Seluwit) 16cm

DESCRIPTION Entirely brilliant, glowing green. Male more emerald, with a yellow spot before eye, black spot behind ear, 3 black wing bars, and more bluish under-tail coverts

Female

and tail. Female more grass-green, lacking markings but with same blue beneath tail. **DISTRIBUTION** Resident from c. 16°N in Myanmar through Peninsular Thailand and Peninsular Malaysia (though now lost from Singapore) to Sumatra, Bunguran and Borneo. **HABITS AND HABITAT** Lowland evergreen rainforest from sea-level to c. 760m, with occasional long-distance dispersal possible at other altitudes. In pairs, singly or in small groups whose significance is unknown (and worth detailed study), snatching fruits and some insects in flight. Nest is a neat hanging bag suspended from an understorey twig, sometimes at human head height. Call is an accelerating, descending series of taps; also short cat-like moans and frog-like croaks.

Whitehead's Broadbill

■ *Calyptomena whiteheadi* (Burung Seluwit Kinabalu) 25cm

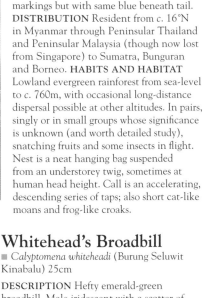

DESCRIPTION Hefty emerald-green broadbill. Male iridescent with a scatter of black marks over breast, black patches on throat and in front of and behind eye, and black bars on wings. Female duller green but still with black throat-patch and obscure mark behind eye. **DISTRIBUTION** Resident only in Borneo, where it is endemic in the mountains. **HABITS AND HABITAT** Occurs rather unpredictably in middle storey of tall lower montane forest and fringes of upper montane forest, at c. 1,000–1500m; rare records of dispersal into lowlands. Fruits are swallowed whole, with regurgitation of astonishingly large, indigestible seeds. Can sit quietly for long periods, but gives characteristic whirring; also gulps and croaks in a social context.

Long-tailed Broadbill
■ *Psarisomus dalhousiae* (Takau Bayan) 26cm

DESCRIPTION Like a parakeet, with a brilliant grass-green back and wings, paler lime-green on breast, and bluer on primaries and long tail. Narrow white collar (perhaps better defined in female) joins yellow throat, with yellow patch over ear; sides and top of head black, with a little blue skullcap. **DISTRIBUTION** Resident from central Himalayan foothills, through S China and Indochina to Peninsular Malaysia, Sumatra and Borneo. Not in Singapore. **HABITS AND HABITAT** Montane forest at *c.* 850–1,500m defines this bird's distribution, with scattered records as low as 250m in foothills. Occurs in small (family?) parties, seeking insects on foliage and twigs of trees in the middle and upper storeys, occasionally in small trees along quiet forest roads. Flock members bob, flaunt tails and call to each other in a descending series of trills.

Silver-breasted Broadbill
■ *Serilophus lunatus* (Takau Hujan) 16cm

DESCRIPTION Most elegant bird, with a silvery-grey head and breast (crossed in female by a silver-white line), shading to an ashy-brown back and rich chestnut rump. Conspicuous black eyebrow; wings black with flashes of blue and white. Bill silvery blue and yellow. **DISTRIBUTION** Resident from E Himalayas through S China and the Indo-Malayan region as far as Peninsular Malaysia and Sumatra (not Singapore or Borneo). **HABITS AND HABITAT** In Peninsular Malaysia occurs only along the Main Range, up to *c.* 1,200m and down to a variable altitude, sometimes as low as 230m, in the middle and lower storeys. Could be considered a hill-slope, rather than strictly montane, bird. Nests are hung from small trees over hillside gullies, the lining replenished with fresh green leaves throughout incubation, as in other broadbills. Thought to feed on insects.

Black-and-red Broadbill

■ *Cymbirhynchus macrorhynchos* (Takau Rakit)
22–24cm

DESCRIPTION Black crown, breast-band, wings and tail; longitudinal white flash on coverts overlying secondaries; large bib above breast-band, breast, belly and rump a rich, deep, gorgeous red. Bill dramatic turquoise and yellow. Sexes are alike. **DISTRIBUTION** Resident in Indochina, Myanmar and Thailand, through Peninsular Malaysia (now only sporadic in Singapore) to Sumatra, Borneo and some intervening islands. **HABITS AND HABITAT** Best viewing opportunities are along forest-fringed rivers, as the untidy dead-leaf nests hang above water from overhead branches or protruding snags. Also seen in mangroves and in forest edge far from water, with some invasion of tall plantations in lowlands, to *c.* 300m. Calls include a rasping, rising trill.

Black-and-yellow Broadbill ■ *Eurylaimus ochromalus* (Takau Kasturi) 15cm

DESCRIPTION Small broadbill with black head, white collar, conspicuous black breast-band (broken medially in female) and pale pink underparts; wings and tail largely black

with multiple flashes of yellow on back, coverts, secondaries and rump. Comical yellow eye and blue bill. **DISTRIBUTION** Resident from central Thailand and Myanmar through Peninsular Malaysia (not, or perhaps no longer, in Singapore), Sumatra, Belitung and Borneo. **HABITS AND HABITAT** In the canopy and middle storey of lowland forest, peat-swamp forest and, sometimes, rubber plantations, although it may be hard to see. Bag-like dead-leaf nest is hung from a branch or twig, usually above a space. Advertises its presence by a long series of notes, accelerating and rising, before ending sharply.

Blue-winged Pitta ■ *Pitta moluccensis* (Burung Pacat Sayap Biru) 20cm

DESCRIPTION Brilliantly particoloured bird with black mask, buff underparts that are red beneath tail, green upperparts and bright blue wings. In flight, wings are bold blue and black with white panels, like those of many pittas and some kingfishers. 2 similar species are the large-billed **Mangrove Pitta** *P. megarhyncha* in Peninsular Malaysia and the chestnut-crowned hillside **Fairy Pitta** *P. nympha* migrant in Borneo. **DISTRIBUTION**

Resident in China and Indochina southwards to N Peninsular Malaysia; migrant to rest of Malay Peninsula, Singapore, Sabah and Sarawak, occasionally reaching as far as Java. **HABITS AND HABITAT** Usually alone, in lowland forest or dense vegetation in plantations or even large gardens, where it hops on the ground, turning leaf litter to seek insects and grubs. Migrates at night; many records are of birds stunned by hitting buildings when disoriented by lights. Call is a 4-note *chew-chew, chew-chew*.

Garnet Pitta ■ *Erythropitta granatina* (Burung Pacat Merah) 15cm

DESCRIPTION Deep, shining blue-black, with scarlet crown, lower breast and belly, and bright blue iridescence on bend of wing; narrow brow light blue. Juvenile is dark

brown all over, without any spots or streaks. **DISTRIBUTION** From *c.* 6°N in Peninsular Malaysia, through E Sumatra and Borneo. Formerly resident in Singapore, now locally extinct. W Sumatran birds separated as **Graceful Pitta** *E. venusta* and Sabah birds as Black-crowned Pitta *E. ussheri*. Formerly resident in Singapore, now locally extinct. **HABITS AND HABITAT** The floor of tall lowland forest to *c.* 200m. Nest is a dome of dead leaves on the ground next to a tree root. Call is a low, pure whistle, 1.5 seconds long except in Sabah, where it is 2 or 3 seconds long.

Blue-headed Pitta ▪ *Hydrornis baudii* (Burung Pacat Kepala Biru) 18cm

DESCRIPTION In male, black sides of face separate brilliant blue cap and brilliant white throat; chestnut brown above; deep, dark blue below; long white wing bar. Female brown above and fawn below, with pale throat, long white wing bar on dark wing, and bluish tail. **DISTRIBUTION** Resident only in Borneo, where it is endemic in the lowlands of Sabah and Sarawak, as well as Brunei and Kalimantan. **HABITS AND HABITAT** Solitary individuals are seen foraging for insects among ground litter in lowland evergreen rainforest, usually below 300m but once reported up to 1,200m. Nest is a ball of leaves lodged by roots on the ground. Usual call is a soft *pwi-wi-wi*, with emphasis on the 1st note.

White-bellied Erpornis
▪ *Erpornis zantholeuca* (Burung Rimba Berjambul Hijau) 13cm

DESCRIPTION Pale grey face, breast and (nearly white) belly; crown with short, erectile crest; back, wings and tail light green; yellow touch to fringes of primaries and yellow beneath tail. Sexes are alike. **DISTRIBUTION** Resident from the western Himalayan foothills through southern China to Taiwan, and through Southeast Asia to Peninsular Malaysia (not Singapore), Sumatra and Borneo. **HABITS AND HABITAT** Formerly considered to be a yuhina, but now not even included among the babblers. Seen singly or in pairs in the canopy and middle storey of lowland evergreen rainforest, from the extreme lowlands to *c.* 900m, and occasionally into lower montane forest to a maximum 1,200m. This foliage-gleaning insectivore is a persistent but still scarce participant of mixed foraging flocks.

White-browed Shrike-vireo

■ *Pteruthius aeralatus* (Burung Rimba Kening Putih) 17cm

DESCRIPTION Both sexes with entirely pearl-white undersides. Crown and mask of male black with white eyebrow; back grey; wings and tail black, inner secondaries ochre and chestnut, tips of primaries white. Female with subdued grey head and back, light olive-green wings and tail. **DISTRIBUTION** Resident from W Himalayan foothills through S China discontinuously to Peninsular Malaysia (not Singapore), Sumatra, Borneo and Java. In Peninsular Malaysia it is confined to the Larut and Main ranges, and various outlying mountains. **HABITS AND HABITAT** From 900m in tall lower montane forest to 2,000m in elfin upper montane vegetation, mainly in the canopy but also at forest edges. Territorial pairs call loudly, nesting early Feb–Jun on a branch in the canopy or middle storey. Insectivorous, known to take caterpillars and other invertebrates.

Female

Black-eared Shrike-vireo ■ *Pteruthius melanotis*

(Burung Rimba Telinga Hitam) 12cm

DESCRIPTION All ages and both sexes have a black line before, around and behind eye, curling down around ear coverts and emphasising white eye-ring; crown and back green; wings blacker, with 2 pale bars. Male has orange throat and yellow underparts; female and young have faintly yellow ear coverts and pearly breast. **DISTRIBUTION** Resident from central Himalayan foothills through S China discontinuously to Peninsular Malaysia (not Singapore). **HABITS AND HABITAT** Confined to tall montane forest of the Main Range and various outlying mountains at *c.* 1,000–1,800m, where it forages in the canopy and middle storey. Insectivorous, taking caterpillars and, presumably, other insects by searching the foliage, lichen-covered branches and trunks. Call is a scolding trill.

Malaysian Cuckooshrike ■ *Coracina larutensis* (Burung Kelabu Gunung) 28cm

DESCRIPTION Previously called *C. novaehollandiae*. Subtly shaded ash-grey all over, darkest on face and around base of bill, palest on belly and beneath tail; robust bill and legs

black. **DISTRIBUTION** Resident from W Himalayas through S China to Taiwan, and Southeast Asia to Peninsular Malaysia (not Singapore, Sabah or Sarawak), Java and Bali. Other cuckooshrike species occur in lowland forest of Peninsular Malaysia, and lowland and montane forest of Sabah and Sarawak. **HABITS AND HABITAT** Singly or in pairs in the canopy of montane forest at 1,000–2,050m, foraging for small, round fruits and arthropods among the foliage, and taking some insects in flight. Nest is a deep cup of twigs and lichen in the horizontal fork of a branch. Has an amusing habit of alternately lifting each folded wing.

Black-winged Flycatcher-shrike
■ *Hemipus hirundinaceus* (Rembah Sayap Hitam) 15cm

DESCRIPTION Male has black face and crown, neck, back, wings and tail; pure white rump; pearly-grey to white underparts from chin to vent. Female is like male but duller

and browner. **DISTRIBUTION** From *c.* 7°N southwards through the Malay Peninsula, Sumatra, Borneo and intervening islands to Java and Bali. Resident in Peninsular Malaysia, Sabah and Sarawak; never reliably recorded from Singapore. **HABITS AND HABITAT** Lowland forest, swamp forest and the landward side of mangroves, from sea-level up to *c.* 300m, rarely to 800m. Lives in the upper storey, or lower at the forest edge. Usually seen as a pair or alone, flycatching and gleaning the foliage for insects. Call is a short, fairly harsh trill.

Pied Triller ▪ *Lalage nigra* (Rembah Kening Putih) 17cm

DESCRIPTION Small birds, black above and white below, with a white brow and extensive white wing panel, and pale grey rump. Female has similar pattern but all markings are more subdued; also slightly scaly on breast. **DISTRIBUTION** Resident in Andaman and Nicobar islands through Peninsular Thailand, Peninsular Malaysia and Singapore to Sumatra, Belitung, Borneo, the Philippines and Java. **HABITS AND HABITAT** In coastal lowlands, from pandan scrub and the landward side of mangroves on the coast, to secondary woodland, parks and large gardens inland, including suburban fringes and isolated trees in e.g. buffalo grazing grounds. Likes to forage in the crowns of tall trees, but also comes down low, seeking insects among the foliage. Nest is a cup of dead fibres and casuarina needles, decorated with lichen, in the horizontal fork of a small branch. Harsh buzzing and complaining sounds, none very loud.

Female

Ashy Minivet ▪ *Pericrocotus divaricatus* (Burung Matahir Kelabu) 19cm

DESCRIPTION Small black and white bird, longer and slimmer than Pied Triller (above), and with short, narrow white brow and long, narrow white wing bar (not a broad panel). Sexes are alike. **DISTRIBUTION** Breeds from Siberia through N China to Korea and Japan; winters southwards to India and Southeast Asia, including Peninsular Malaysia, Singapore, Sabah, Sarawak and the Philippines. **HABITS AND HABITAT** Non-breeding birds are characteristic of coastal forest, but also occur in the canopy of lowland and peat-swamp forest, and occasionally up into lower montane forest. Flocks of 20 or more trickle from tree to tree, follow-my-leader fashion, calling soft, tinkling notes and foraging for insects before moving on.

Grey-chinned Minivet ■ *Pericrocotus solaris* (Burung Matahari Gunung) 18cm

DESCRIPTION Adult male black and red, chin only slightly greyer than head; single long red wing bar (male Scarlet Minivet, below, has a 2nd red wing spot). Female has similar pattern, but is dark grey and yellow, with dark grey forehead. Similar **Fiery Minivet** *P. igneus* is slightly smaller with a different flight call (an upward, slurred *swee-eet*).
DISTRIBUTION Resident from central Himalayan foothills through S China to Taiwan, and in highlands southwards to Peninsular Malaysia (not Singapore), Sumatra and Borneo, including Sabah and Sarawak. **HABITS AND HABITAT** In the crowns of tall trees in lower and upper montane forest, 975–2,075m, into stunted elfin forest. Searches for insects in foliage while perched, and hovers at shoot tips; in pairs and, outside the Feb–Apr breeding season, in flocks of up to 30. Flight call is a twittering *sri-sisi*, repeatedly.

Male   *Female*

Scarlet Minivet ■ *Pericrocotus flammeus* (Burung Matahari Besar) 19cm

DESCRIPTION Male black and red, very like Grey-chinned Minivet (above) but with a 2nd small, rounded patch of red on secondaries. Female grey and yellow, paler than female Grey-chinned, with 2nd (yellow) wing-patch and yellowish forehead. DISTRIBUTION Resident from the W Himalayas and Indian sub-continent through S China and Southeast Asia to Peninsular Malaysia, Singapore, Sumatra, Belitung, Borneo, the Philippines, Java, Bali and Lombok. **HABITS AND HABITAT** Like other minivets, it prefers the crowns of tall trees, in this case in lowland evergreen rainforest from the extreme lowlands to *c.* 1,000m (just into lower montane forest, where it overlaps with Grey-chinned), and in peat-swamp forest. All sorts of small invertebrates are taken; pairs after breeding (Jan–Jun) group into flocks.

Bornean Whistler
▪ *Pachycephala hypoxantha* (Murai Mas) 16cm

DESCRIPTION Sides of face and entire underparts from chin to vent bright yellow; lores dark, upperparts olive-green without wing bars. Females duller than males on face and underparts. **DISTRIBUTION** Resident and endemic in montane Borneo, including Sabah, Sarawak and Kalimantan. **HABITS AND HABITAT** Middle storey of tall montane forest, at 830–2,400m, but commonest at c. 1,300–1,800m. Will come to quiet forest edges, sometimes near streams and in montane valleys, flycatching and taking insects from twigs and foliage. Usually seen singly or in mixed-species flocks, with breeding known or suspected Nov–Mar.

Mangrove Whistler ▪ *Pachycephala cinerea* (Murai Bakau) 16cm

DESCRIPTION Entirely ashy brown, palest on the belly, without any wing bars or eye-ring. Eye and the robust bill black; legs are grey. **DISTRIBUTION** Resident from E India to Indochina and Southeast Asia, as far as Peninsular Malaysia, Singapore, Sumatra, Borneo, Java and Lombok. **HABITS AND HABITAT** Intrudes from mangroves into peat-swamp and freshwater-swamp forest, and recorded occasionally inland in forest over impoverished soils in Sabah and Sarawak. An insectivore, with breeding noted in Mar–Jun. Though plain, it is a common and easily located bird in mangroves because of its song: 4 or 5 staccato whistles on 1 pitch, followed by a couple of higher-pitched notes and a whistled whip-crack ending.

Black-naped Oriole ■ *Oriolus chinensis* (Burung Kunyit Besar) 26cm

DESCRIPTION Males brilliant yellow over most of plumage – even black wings and tail have yellow bars and flashes. Black band runs from bill through eye, joining at back of head. Bill rosy pink, feet grey. Females slightly duller, more olive above. Juveniles olive-green, streaky below, with only faint indications of future adult pattern. **DISTRIBUTION** Naturally resident from Mongolia discontinuously through E and Southeast Asia to

the Lesser Sundas. Also migrant to much of Southeast Asia, including Peninsular Malaysia, Singapore and, rarely, Borneo; natural and human-assisted spread of residents has occurred through Singapore and Peninsular Malaysia since 1925. **HABITS AND HABITAT** Gardens, parkland, orchards and secondary woodland in lowlands are its preferred habitat, where it forages for all sorts of fruits and insects; also mangroves. There is much interaction between individuals, with loud, fluting calls *ku-eyou-ou*, instantly recognisable but tremendously variable, and chasing, following and displacement from food sources.

Black-hooded Oriole ■ *Oriolus xanthornus* (Burung Kunyit Topeng Hitam) 23cm

DESCRIPTION Adult brilliant yellow with black head and upper breast, and black tail and wings with yellow markings. Young birds are duller, with less intense black on wings, tail, crown and ear coverts, and streaked with whitish on throat. **DISTRIBUTION** From the Himalayan foothills to Sri Lanka and SW China, through Indochina, discontinuously in Thailand, and to N Sumatra and NE Borneo. Resident in Peninsular Malaysia perhaps only in Langkawi, and in a very limited area in E Sabah. Scarce migrant in Peninsular Malaysia southwards to Fraser's Hill. **HABITS AND HABITAT** In mangroves, the landward side of mangroves and adjacent wooded landscapes; an apparently relict distribution is confined to the drier seasonal corners of the Sunda region. Records of migrants up to *c.* 1,300m. Little local information is available.

Black-and-crimson Oriole

■ *Oriolus cruentus* (Murai Hitam Merah) 22cm

DESCRIPTION Entirely black, except male has a brilliant crimson breast-patch and crimson primary wing coverts; bill silvery grey, feet grey. In females, breast-patch may be faintly indicated by grey tone, but Borneo females are said to develop red patch as brilliant as that of males – this confusion now being resolved. **DISTRIBUTION** Resident in the mountains of Peninsular Malaysia, Sumatra, Borneo and Java. **HABITS AND HABITAT** Singly or in pairs in tall forest from 600m on hill slopes to 1,500m in montane forest in the Malay Peninsula, and to 2,300m in Borneo. Keeps largely to the canopy, but sometimes in the middle storey and edge along quiet forested roadsides or tracks, taking foliage-eating caterpillars as well as other insects and some fruits. A cat-like mewing and harsh nasal notes have been described throughout its range; a melodious call is mentioned only from Borneo.

White-breasted Woodswallow

■ *Artamus leucorynchus* (Tirjup Layang-layang Dada Putih) 17cm

DESCRIPTION Smooth grey above and white below and on rump; head and bib darker grey. Tail slightly forked. In flight, wings appear triangular and starling shaped, but birds glide for great distances between bouts of flapping. Juveniles have buff fringes to wing feathers. **DISTRIBUTION** Resident in the Andamans, Sumatra, Borneo, the Philippines, Java and Bali eastwards to Pacific, including Fiji and the New Hebrides. Abundant in Sabah and Sarawak; first colonised the W coast of Peninsular Malaysia in 1977, and still scarce and restricted there. **HABITS AND HABITAT** The flight behaviour makes even distant birds easy to identify, and the bill can be heard snapping as they catch flying insects. Perched birds are sociable but boisterous to aggressive, performing mutual preening and huddling in social roosts.

Female

Common Iora
■ *Aegithina tiphia* (Burung Kunyit Kecil) 13cm

DESCRIPTION Green above with black tail and, in male, black on crown; yellow below, fading to white on under-tail coverts, which curl sideways and up over rump to make rump seem white; wings black with 2 white bars and pale fringes. **DISTRIBUTION** Resident from foothills of the W Himalayas to India, Sri Lanka, S China, Peninsular Malaysia, Singapore, Sumatra and its outlying islands, Borneo, Java, Bali and the Philippines. **HABITS AND HABITAT** In display flight from tree to tree, male sings and shows off his false white rump. Takes insects from foliage in mangroves and the landward side of mangroves, plantations, trees in parks and roadsides, and at the edge of lowland evergreen forest. Often forages high in the canopy, for example in tall *Albizia* trees, but also comes down to the understorey. Breeds Jan–Jun. Has a huge range of calls and song types.

Green Iora ■ *Aegithina viridissima*
(Burung Kunyit Hijau) 13cm

DESCRIPTION Deep sage-green, with a cream mark above and below eye, and wings blackish with 2 pale bars and pale fringes. The pale under-tail coverts curl up over rump to make it appear at least partly white. Female duller than male with lighter tail and paler head. **DISTRIBUTION** Resident from *c.* 13°N in Thailand and 12°N in Myanmar through Peninsular Malaysia, Singapore, Sumatra, the Natunas and Borneo. **HABITS AND HABITAT** An attractive, small bird of the canopy of lowland evergreen rainforest from sea-level to *c.* 820m, peat-swamp forest and tree plantations; also within mangroves in Sabah and Sarawak, and on some islands off Peninsular Malaysia. In pairs or small groups, seeking all kinds of soft-bodied insects among foliage, and also taking some small soft fruits such as figs; often a participant in mixed foraging flocks in the canopy.

White-throated Fantail ■ *Rhipidura albicollis* (Murai Gila Gunung) 19cm

DESCRIPTION Blackish above and below, except for the white throat triangle, white brow and white tips to tail feathers. Juveniles duller and browner, with less well defined brow and throat mark. **DISTRIBUTION** Resident from the W Himalayas to S China and Southeast Asia as far as Peninsular Malaysia (not Singapore), Sumatra and Borneo. **HABITS AND HABITAT** Confined to lower montane forest, usually from *c.* 850m upwards into upper montane forest as high as 2,070m. Seeks insects in the middle and lower storeys, where it is one of the commoner birds and a common participant in mixed foraging flocks. Nest is a cup slung in the fork of a small lateral branch in the middle storey, hardly big enough to contain 2 growing young. Has a repeated tuneful song of 7 or 8 notes

Sunda Pied Fantail ■ *Rhipidura javanica* (Murai Gila Biasa) 18cm

DESCRIPTION Black above and white below (throat, lower breast and belly) with a broad black breast-band; flaunted tail is black with white tips. Adults have a short, narrow white brow, this obscure in the browner juveniles. **DISTRIBUTION** Resident from S Indochina through Peninsular Thailand and Peninsular Malaysia to Singapore, Sumatra, Belitung, Borneo, the Philippines, Java and Bali. **HABITS AND HABITAT** Common in most tall mangroves, extending into peat-swamp and other wet forested habitats, forest edges, plantations, gardens and secondary growth on abandoned land. This and other fantails skip and flirt through the vegetation, swinging their body this way and that, and fanning the tail repeatedly, to disturb insects that are then snapped up.

Ashy Drongo ■ *Dicrurus leucophaeus* (Cecawi Kelabu) 28cm

DESCRIPTION Several subspecies occur. Those resident on Borneo mountains are always pale grey with whitish sides to face and red eye; in Malay Peninsula and Singapore, similar-looking migrants occur, but also nearly black glossy residents. All have ruby eye and deeply forked, faintly upswept tail. Migrant **Black Drongo** *D. macrocercus* is elegant with a dark eye and more upswept tail tips; migrant **Crow-billed Drongo** *D. annectans*

is stocky with a heavier bill and spangled plumage. DISTRIBUTION Resident from Afghanistan eastwards through S China and Southeast Asia to Palawan, Borneo, Java, Bali and Lombok. Migrant to Southeast Asia, including Peninsular Malaysia and Singapore. HABITS AND HABITAT Predominantly in mangroves, and open ground with fairly closely spaced trees that provide sites for foraging on insects. Migrants tend to occur more in parkland and gardens, but in Sabah and Sarawak they live in the forest canopy at 500–2,200m in hill and montane forest. Mimics other bird calls, but less well than the Greater Racket-tailed Drongo (below).

Malay Peninsula and Singapore (left); Borneo (right)

Greater Racket-tailed Drongo
■ *Dicrurus paradiseus* (Cecawi Kera) 32–57cm

DESCRIPTION Noisy, conspicuous, glossy black bird with red eye; the 2 outer-tail feathers project as wires, with 1 rounded and twisted racket on each side, 1 or both of which may be missing owing to moult or damage. DISTRIBUTION Resident from the Indian sub-continent eastwards to Hainan and through Peninsular Malaysia, Singapore, Sumatra, Borneo, Java and Bali. HABITS AND HABITAT The canopy and middle storey of lowland evergreen rainforest from the extreme lowlands to *c.* 850m, and in mangroves, tree plantations (rubber and oil palm), parkland with abundant trees, and secondary woodland. Typically seen in pairs, and flies out from a high perch to catch passing insects. Gives a wide range of calls, including good imitations of many other birds.

Black-naped Monarch ■ *Hypothymis azurea* (Kelicap Ranting) 16cm

DESCRIPTION Bright blue forequarters, shading down to ashy-brown wings and tail, and whitish belly. Male has black throat bar, crown spot and dab over bill. Female has greyer breast and black dab over bill.

DISTRIBUTION Resident from India through S China and Southeast Asia to Peninsular Malaysia (and formerly Singapore), Borneo, the Philippines and the Lesser Sundas. **HABITS AND HABITAT** In the lower and middle storeys of lowland evergreen rainforest, from the extreme lowlands to *c.* 915m in the Malay Peninsula, or 1,100–1,200m in Sabah and Sarawak. Usually alone or in pairs, or in mixed foraging flocks. One of the commoner birds glimpsed during journeys along forested rivers, aerial flycatching within the forest and occasionally bathing by diving into the water.

Blyth's Paradise-flycatcher
■ *Terpsiphone affinis* (Murai Ekor Gading) 22–40cm

DESCRIPTION Female has black cowl shading down to grey neck and breast, and whitish belly; rufous-chestnut wings and tail. Male either similar and with an extremely long tail (nearly triple its body length), or else body, wings and tail are pure white with fine black edgings to some feathers. Significance of the male colour types is unclear, but it may not be age related. **DISTRIBUTION** Resident from Indochina southwards to Peninsular Malaysia, Sumatra, Java and formerly Singapore. Non-breeding migrant **Amur Paradise-flycatcher** *T. incei* also occurs. **HABITS AND HABITAT** The white-phase male is spectacular, and is not rare as a resident in the middle and upper storeys of tall lowland forest. Migrants also occur in tall secondary woodland. It is usually solitary and insectivorous, taking fairly big insects, and lives from the extreme lowlands to *c.* 880m in the Malay Peninsula, and up to 1,200m in Sabah and Sarawak.

House Crow
■ *Corvus splendens* (Gagak Rumah) 42cm

DESCRIPTION Typical crow, glossy black all over, except that the black face and forehead are defined rearwards by a paler zone of grey over the hind-face and neck, down to sides of breast. This colour develops with age, juveniles looking entirely black. **DISTRIBUTION** Naturally resident from Afghanistan eastwards through Indian sub-continent and S China to SW Thailand. Widely introduced by man as far as S Africa, Europe, North America and Australia. Present in Peninsular Malaysia and Singapore since *c.* 1904, and a few birds in Kota Kinabalu, Sabah, since 1998. **HABITS AND HABITAT** The abundant crow of Singapore and Peninsular Malaysian towns (at least, the larger towns on the W coast), forming noisy communal roosts and mainly dependent on man-made refuse.

Bornean Green Magpie
■ *Cissa jefferyi* (Gagak Hijau Ekor Pendek) 35cm

DESCRIPTION Brilliant lime-green bird the size of a small crow, with a black face mask, chestnut wings, and white tips to tail and innerwing feathers. The **Common Green Magpie** *C. chinensis*, found lower down in the mountains of Borneo and Peninsular Malaysia, is very similar but has a longer tail, black and white (not just white) tips to innerwing feathers, broader black subterminal bars on tail, and sometimes traces of yellow on crown. Both species have red bill, legs and skin around eye, but this is dark red lake in Short-tailed and brilliant red in Common Green. **DISTRIBUTION** Resident in mountains of Borneo. Distinct from Asian continental **Green Magpie** *C. hypoleuca* and **Javan** *C. thalassina*. **HABITS AND HABITAT** In Sabah and Sarawak this bird replaces the Common Green at *c.* 900m, and in lower and upper montane forest reaches 2,440m. Although the 2 species' altitudinal ranges overlap, they never seem to occur together. A range of whistles and buzzing notes may reveal a solitary bird or small group in dense vegetation of the middle or lower storey, seeking beetles, caterpillars, snails and other invertebrates.

Bornean Treepie

■ *Dendrocitta cinerascens* (Burung Tontihak) 45–50cm

DESCRIPTION Ashy brown from forehead and over face to breast, shading more rufous towards belly; crown and back grey, wings black with basal white patch, and very long tail grey with black tip. Sexes are similar, juveniles only a little duller. **DISTRIBUTION** Resident in the mountains and hill slopes of Borneo. Now considered distinct from the closely related **Sumatran Treepie** (*D. occipitalis*). **HABITS AND HABITAT** In the canopy and middle storey of montane forest, to a maximum 2,900m; usually not below 800m, but occasional records on hill slopes down to 100m. Attractive bell-like calls are interspersed with buzzes, made when perched and in flight, and answered between individuals, pairs and, occasionally, small flocks. Frugivore and insectivore, taking most food from the canopy but even coming to the ground in forest clearings. Breeding has been reported in Jan–Mar.

Bornean Black Magpie ■ *Platysmurus aterrimus* (Burung Kambing) 40cm

DESCRIPTION Heavy, active black bird, tail almost as long as head and body, with a slight crest and red iris. In Peninsular **Malaysian** *P. leucopterus* white wing-patch is visible from coverts to secondaries, but endemic **Bornean** *P. aterrimus* (shown) is entirely black. **DISTRIBUTION** Resident from *c.* 14°N in Myanmar and Peninsular Thailand through Peninsular Malaysia, Bintan, Sumatra, Bangka and Borneo, including Sabah and Sarawak. **HABITS AND HABITAT** Typically in the upper and middle storeys of lowland evergreen rainforest, occasionally peat-swamp forest, from the extreme lowlands to *c.* 300m, and in scraps of gallery forest and adjacent tree plantations. Usual food includes insects and small vertebrates, but some fruit too. Bell-like hoots or a sheep-like bleat, separately or interspersed, are made between individuals, which forage in pairs or small social groups.

ay Shrike ■ *Platylophus galericulatus* (Burung Menjerit) 32cm

DESCRIPTION Blackish or, in the Borneo race photographed, chocolate-brown with a fine vertical or forward-tilting crest; white marks around eye and white patch on side of neck.

DISTRIBUTION Resident from *c.* 13°N in Peninsular Thailand through Peninsular Malaysia (but not Singapore) to Sumatra, Borneo and Java. **HABITS AND HABITAT** Not easy to see, but can be curious, raising a shrike-like rattling chatter when it sees people. Keeps to the middle and lower storeys of lowland evergreen rainforest, rarely in montane forest to 1,200m in Peninsular Malaysia and a record 1,525m in Sarawak. Usually solitary or in pairs, seeking invertebrates such as beetles, grasshoppers, wasps and cicadas. Nest is a sturdy cup made of twigs on a lower branch, and has only been described recently, from a single observation in Java.

Tiger Shrike ■ *Lanius tigrinus* (Tirjup Harimau) 18cm

DESCRIPTION Back, wings and tail rich brown; back and lesser wing coverts always finely barred black. Strong contrast between brown back and grey cap; black mask and relatively thick bill. Females may show some barring on flanks. **DISTRIBUTION** Resident in far E Russia, Japan and Korea through central China; non-breeding migrant to S China and Southeast Asia as far as Java and Bali. **HABITS AND HABITAT** In forest edges, bamboo groves, dense roadside vegetation and abandoned cultivation, extending into heavily logged forest. Noted singly throughout Peninsular Malaysia, Singapore, Sabah and Sarawak to *c.* 850m, but most Borneo records are only from the N. Beetles and grasshoppers are the chief prey.

Brown Shrike ■ *Lanius cristatus* (Tirjup Padang) 19cm

DESCRIPTION Several subspecies occur, these differing in tone but always with a plain back, never finely barred. Black mask, pale supercilium and either brown or grey crown; back and tail brown, undersides off-white; bill relatively small. Sexes are alike.
DISTRIBUTION Resident through vast areas of temperate E Asia, from 70°E to Sakhalin and southwards through China; migrates to China, India and Southeast Asia as far as Sumatra, Borneo, the Philippines, Java and the Lesser Sundas.

HABITS AND HABITAT In our non-breeding range, arriving birds set up territories in cultivation, gardens, and open ground with bushes, scattered trees and grassland, advertising their presence with a chattering call around their Sep arrival and Mar departure. They occur singly, on fences, trees and bush-tops, descending to catch invertebrates (mostly beetles and grasshoppers) on the ground, but occasionally lizards and small birds.

Long-tailed Shrike ■ *Lanius schach* (Tirjup Ekor Panjang) 26cm

DESCRIPTION Black mask extending over forehead; underparts white with peach-buff flanks; crown and back grey, merging to buff on scapulars. Long tail and blackish wings showing strong contrast with body; white wing-spot, most visible in flight.

DISTRIBUTION Resident from Kyrgyzstan through central Asia and the Indian sub-continent to all but N China, and patchily through Southeast Asia to Java, Bali, the Lesser Sundas and New Guinea. N populations migrate, a few as far as Borneo. **HABITS AND HABITAT** Spread of cultivation suggests possible arrival of this species in Peninsular Malaysia in the 2nd half of the 19th century. Still found only W of the Main Range but commonly in Singapore; in Sabah and Sarawak perhaps just a scarce migrant. In open rice fields and grassland with shrubs, taking mainly insects.

Brown-throated Sunbird ■ *Anthreptes malacensis* (Kelicap Mayang Kelapa)
13cm

DESCRIPTION Male has brilliant yellow breast and belly; throat glossy purplish brown; crown purple, shading to blue wings and tail; sides of face and wings olive-green. Female olive-green, fairly bright yellow on breast and belly, and with yellow throat contrasting with

sides of face. **DISTRIBUTION** Resident from Myanmar and central Thailand through Indochina, Peninsular Malaysia, Singapore, Sumatra, Borneo and the Philippines to Sulawesi, Java, the Lesser Sundas and most intervening small island groups. **HABITS AND HABITAT** 1 of the 2 most common sunbirds. Seen singly or in pairs in parks and gardens, coconut and other tree plantations, secondary woodland and mangroves. Feeds on much nectar (e.g. from coconut palms, and hibiscus and tubular flowers), small colourful fruits, and invertebrates such as spiders. Usually in the lowlands but recorded up to 900m, with breeding mostly Jan–Aug.

Red-throated Sunbird ■ *Anthreptes rhodolaemus* (Kelicap Pinang) 13cm
DESCRIPTION Care is needed to distinguish it from Brown-throated Sunbird (above).

Male has maroon-red sides of face and upperwing coverts, with only a small blue carpal wing-patch; crown and upperparts iridescent greenish. Female like female Brown-throated but with duller greenish sides to breast. **DISTRIBUTION** Patchily from *c*. 12°N in the Malay Peninsula, southwards through Sumatra and Borneo. Resident in Peninsular Malaysia, Sabah and Sarawak; presumed former resident in Singapore, now locally extinct. **HABITS AND HABITAT** The upper storey of tall lowland forest, including lightly disturbed forest and peat-swamp forest, up to *c*. 400m or, rarely, 900m. Insects, nectar and fruit are taken. It may be less territorial, or at least less aggressive, than Brown-throated.

Ornate Sunbird ■ *Cinnyris ornatus* (Kelicap Pantai) 12cm

DESCRIPTION Olive-green from crown to rump, with lower breast and belly bright, deep yellow. In male, forehead, throat and upper breast are iridescent blue-black; female is paler and duller overall with whitish chin, and lacks any blue-black markings.
DISTRIBUTION Resident from mainland Asia through to Lesser Sundas, including Malaysia, Singapore, Sabah and Sarawak.
HABITS AND HABITAT Locally the most common and conspicuous sunbird throughout the region, in parks and gardens, scrub, mangroves and plantations. Takes nectar from a wide range of flower species and shapes, including mangrove trees as well as garden flowers, plus many invertebrates, especially spiders. Tiny but very active, often in pairs, with territorial males pursuing each other from treetop to treetop.

Crimson Sunbird ■ *Aethopyga siparaja* (Kelicap Sepah Raja) 11–13.5cm

DESCRIPTION Deep scarlet red with a yellow rump and dark purple tail. Lower breast and belly grey, and sides of face dark with violet moustache stripe; iridescent green forehead and forecrown (often looking black). Wings, including wing coverts, black. Female olive-grey, darker olive on breast than female Temminck's Sunbird (p. 100) and without any rufous in tail. **DISTRIBUTION** Resident from the central Himalayan foothills through S China and throughout all Southeast Asia to the Philippines and Sulawesi. **HABITS AND HABITAT** Nectar and small invertebrates are the main foods, *Heliconia* flowers being favourites. Similar to, and the open-edge equivalent of, Temminck's Sunbird in the lowlands, occurring in forest edges, gardens, mangroves and peat-swamp forest. There are both habitat and altitudinal trade-offs between Temminck's and Crimson in Borneo, and between these and Black-throated Sunbird (p. 100) in the Malay Peninsula.

Temminck's Sunbird ■ *Aethopyga temminckii* (Kelicap Merah) 10–12.5cm

DESCRIPTION Brilliant scarlet overall with yellow rump and scarlet tail. Lower breast and belly off-white; on face, 2 violet moustache stripes and 2 crown stripes join over nape. Black wings with red coverts. Female olive-grey with rufous sides to base of tail, and faintly rufous fringes to wing feathers; rump plain. **DISTRIBUTION** Resident from *c.* 8°30'N in Peninsular Thailand and Malaysia (but not Singapore) through to Sumatra and Borneo. **HABITS AND HABITAT** The middle and lower storeys of lowland evergreen rainforest, from the extreme lowlands to 300m, and exceptionally to 1,200m in the Malay Peninsula and 1,650m in Sarawak and

Sabah. Usually solitary or in pairs, taking nectar from a variety of forest epiphytes, rhododendrons and introduced flowers, plus various small insects. Formerly called the Scarlet Sunbird, but that name is now reserved for *A. mystacalis* in Java.

LEFT: *Male*. ABOVE: *Female*

Black-throated Sunbird

■ *Aethopyga saturata* (Kelicap Gunung) Male 15cm; female 11cm

DESCRIPTION Male overall very dark with a yellow rump and long central tail feathers; head iridescent blue-black, breast and back maroon, belly grey. Female olive-grey with a

pale yellow rump and grey throat. **DISTRIBUTION** Resident from the central Himalayas through S China and discontinuously in Southeast Asian highlands to Peninsular Malaysia. **HABITS AND HABITAT** Found at 820–2,000m, from the canopy to the lower storey and edges of lower and upper montane forest, including stunted forest on ridgetops. Seen singly or in pairs, often in mixed foraging flocks, taking tiny invertebrates, and nectar from the tubular flowers of forest epiphytes as well as from garden flowers and weeds. Breeding is inferred over a wide range of months.

Streaked Spiderhunter ■ *Arachnothera magna* (Kelicap Jantung Gunung) 18cm

DESCRIPTION Olive-green above from forehead to tail, and buffy white below from chin to vent, entire plumage finely streaked blackish. Bright orange-yellow feet often clearly visible. **DISTRIBUTION** Resident from central Himalayan foothills through S China and the highlands of Southeast Asia to Peninsular Malaysia. **HABITS AND HABITAT** A characteristic bird of lower and upper montane forest in the Malay Peninsula at *c.* 800–1,800m, in the middle and upper storeys, and in roadside vegetation at the hill stations. Takes nectar from banana flowers at the forest edge, and insects from the tangles of epiphytes and lichen on branches. Nesting is remarkably under-recorded considering the species is fairly common. Quick, single alarm notes and a 2-note flight call can often be heard.

Little Spiderhunter ■ *Arachnothera longirostra* (Kelicap Jantung Kecil) 16cm

DESCRIPTION Face and upper breast grey, shading to yellow belly; narrow, dark moustache borders pale sides of face; crown is scaly dark grey, leading to olive wings and tail. Orange tuft may be revealed at bend of wing. **DISTRIBUTION** Resident in the Himalayas and India, through S China and Southeast Asia to Peninsular Malaysia, Singapore, Sumatra, Borneo, Java, and central and S Philippines. **HABITS AND HABITAT** Among the commonest birds of the understorey in logged and unlogged lowland evergreen rainforest, from the extreme lowlands into montane forest at 1,680m. Typically associated with wild bananas, whose flowers are a major source of nectar; also eats invertebrates. Lively and often noisy, with an endlessly repeated *chip* when perched, or singly when in flight.

Yellow-eared Spiderhunter

■ *Arachnothera chrysogenys* (Kelicap Jantung Telinga Kuning) 18cm

DESCRIPTION Dark olive-green spiderhunter, with breast and belly faintly streaked, reaching to yellow thighs; bright yellow but often incomplete eye-ring, which typically touches large yellow cheek-patch. Juveniles are duller, especially eye-ring and cheek-patch. Very similar **Spectacled Spiderhunter** *A. flavigaster* is slightly larger and has a complete yellow eye-ring separated from yellow on ears. **DISTRIBUTION** Resident from *c.* 13°N in Thailand through Peninsular Malaysia and Singapore to Sumatra, Riau, Borneo and Java. **HABITS AND HABITAT** In lowland evergreen rainforest, from sea-level upwards, and at least visiting montane forest to a maximum record of 2,010m. Also in logged forest and tree plantations, and visits roadside trees such as *Erythrina*. Flowers of epiphytes and canopy trees provide most of its food. Call is a single loud *chak!*

Whitehead's Spiderhunter ■

Arachnothera juliae (Kelicap Jantung Tasam) 18cm

DESCRIPTION Blackish brown all over, except that all feathers of head and body have white streaks, these becoming larger and more conspicuous on breast and flanks; wings and tail plain blackish, and rump and under-tail coverts brilliant yellow. **DISTRIBUTION** Endemic to the mountains of Borneo, where it occurs in limited parts of Sabah, E Sarawak and East Kalimantan. **HABITS AND HABITAT** Lower and upper montane forest at *c.* 950–2,100m, where it is found singly or in pairs in the canopy of primary and tall secondary forest. The diet, behaviour and nesting are hardly described. Not very conspicuous, but has a wide range of calls, including a 2-note buzzing *wee-chit*, and an array of chattering and trilling sounds.

Yellow-breasted Flowerpecker
■ *Prionochilus maculatus* (Sepah Puteri Raja) 9cm

DESCRIPTION Dark olive-green above. White moustache and white chin separated by dark malar stripe; rest of underparts bright yellow with strong olive-green streaks, leaving a central band of unmarked yellow down breast. Inconspicuous crown spot is fiery orange in male, dull ochre in female. **DISTRIBUTION** Resident from *c.* 13°N in Peninsular Thailand through Malaysia to Singapore (where it is a past resident and possible dispersant), Sumatra, Bunguran and Borneo. **HABITS AND HABITAT** Common in the middle and lower storeys of lowland evergreen rainforest from the extreme lowlands to *c.* 900m, and sparsely in lower montane forest to 1,250m in Sabah and Sarawak, and up to 1,500m in Peninsular Malaysia. An arboreal foliage-gleaning insectivore and partial frugivore, usually seen alone.

Orange-bellied Flowerpecker
■ *Dicaeum trigonostigma* (Sepah Puteri Dada Biru) 8cm

DESCRIPTION Male with slaty-blue head, upper breast, back, wings and tail; grey throat; brilliant orange from lower breast to vent, and orange-yellow lower back and rump. Female olive-grey, unstreaked, with creamy-yellow rump and centre to belly. **DISTRIBUTION**
Resident from Bangladesh through Peninsular Thailand and Malaysia to Singapore, Sumatra, Borneo, the Philippines, Java and Bali. **HABITS AND HABITAT** Edges of lowland evergreen rainforest from sea-level upwards into montane habitats, with a typical maximum of 1,200m (exceptionally 1,650m in Sarawak). Also enters tall plantations, well-wooded parkland, logged forest and, occasionally, mangroves and peat-swamp forest. Feeds on varied small fruits, including mistletoes, plus insects and nectar. Nest is a hanging pouch in the understorey, with records over a wide scatter of months.

Bornean Flowerpecker ▪ *Dicaeum monticolum* (Sepah Puteri Kerongkong Merah) 8cm

DESCRIPTION Male has a dark grey head, shading to dark blue on wings and tail, more buff on belly; throat and upper breast with a brilliant carmine patch. Female similar, olive-grey with less blue on wings and lacking red patch. **DISTRIBUTION** Resident and endemic in mountains of Borneo (Sabah, the E part of Sarawak, and Kalimantan), though some books include this within the species **Grey-sided Flowerpecker** *D. celebicum* of Sulawesi. **HABITS AND HABITAT** Recorded down to 500m, but typically in the middle and upper storeys of montane forest at *c.* 1,000–2,100m. Single birds or occasionally pairs seek food among the lichens and epiphytes on branches and trunks, taking small, soft fruits and various insects. Nest is a hanging pouch of plant fibres and moss, decorated outside with lichens, and has been found Nov–Feb.

Scarlet-backed Flowerpecker ▪ *Dicaeum cruentatum* (Sepah Puteri Merah) 8cm

DESCRIPTION Male has a broad scarlet line all the way from forehead over crown and down back to rump; white band extends from throat down centre of breast to vent; black sides of face, wings and tail; grey flanks. Female olive-grey, with a broader creamy band from throat to vent, and red rump. **DISTRIBUTION** Resident from the E Himalayas through S China and Southeast Asia to Peninsular Malaysia, Singapore, Sumatra, Riau and Borneo. **HABITS AND HABITAT** One of the most often seen flowerpeckers, in tall secondary growth, orchards, tree plantations and parkland, usually singly; also along the edges of lowland evergreen and peat-swamp forests to a maximum 870m. Eats soft fruits (nibbled to pieces if too big to swallow), mistletoes and soft-bodied invertebrates. Breeding noted from Nov–Aug.

Male   *Female*

Greater Green Leafbird ■ *Chloropsis sonnerati* (Burung Daun Besar) 21cm

DESCRIPTION Male bright grass-green all over, with a black throat-patch reaching eye, where eyelid forms entirely black surround; superimposed blue moustache. Female has yellow throat, light blue moustache, and yellow ring around eye. Juvenile has yellow throat and separate yellow moustache. **DISTRIBUTION** Resident from *c.* 15°N in Myanmar through Peninsular Thailand and Malaysia to Singapore, Sumatra, the Natunas, Borneo and Java. **HABITS AND HABITAT** In lowland evergreen rainforest, peat-swamp forest and forest edges, spilling out to adjacent parkland. Feeds on soft fruits of secondary growth at the forest edge, figs and, possibly, also nectar from flowers. Usually solitary or in pairs, sometimes joining mixed foraging flocks seeking insects. Song loud and quite attractive, imitating other species.

Male

Female

Lesser Green Leafbird ■ *Chloropsis cyanopogon* (Burung Daun Kecil) 18cm

DESCRIPTION Tough to distinguish. Male's throat-patch tends to be outlined faintly yellow, its bill is proportionately smaller than that of Greater Green Leafbird (above), and black does not completely surround eye. Female has a green throat, blue moustache and no blue on wing. **DISTRIBUTION** Resident from nearly 12°N in Myanmar through Peninsular Thailand and Malaysia to Singapore, Sumatra and Borneo. **HABITS AND HABITAT** Very like that of Greater Green Leafbird, being found in the crowns of trees in lowland evergreen rainforest, secondary forest and forest edges. Also about as common as that species, ranging from the extreme lowlands just into montane forest at *c.* 1,100m. Takes various fruits and figs, possibly nectar, and invertebrates while in mixed foraging flocks. Little is known of its nesting. Has an attractive warbling song.

Blue-winged Leafbird ■ *Chloropsis moluccensis*
(Burung Daun Sayap Biru) 18cm

DESCRIPTION Bright grass-green, the male with a black throat-patch, yellow flush over most of head, blue flash on carpel and down edge of wing, and bluish tail. Female has a green throat and blue moustache like female Lesser Green Leafbird (p. 105), but distinct blue flash and edge of wing. **DISTRIBUTION** Resident from NE India to S China and Southeast Asia, as far as Peninsular Malaysia, Singapore (at least some introduced), Sumatra, the Natunas, Borneo and Java. **HABITS AND HABITAT** Edges of lowland evergreen rainforest, and sometimes also the canopy, as well as peat-swamp forest and secondary woodland, from extreme lowlands to *c.* 1,250m. In Borneo mountains and at all elevations in Sabah, it is replaced by the Bornean Leafbird *C. kinabaluensis*, which in both sexes closely resembles male Blue-winged. A very wide range of fruits and invertebrates is eaten. There is little information on its breeding or song.

Male *Female*

Orange-bellied Leafbird ■ *Chloropsis hardwickii* (Burung Daun Bukit) 19cm

DESCRIPTION Bright sage-green above and subtle orange below, with black sides to face and throat, and a long purplish-black panel along wing. Female grass-green with an orange flush on lower belly and under-tail coverts; limited blue on wing coverts and inner secondaries. **DISTRIBUTION** Resident from W Himalayan foothills through S China to Hainan, and in Southeast Asia to the mountains of Peninsular Malaysia. **HABITS AND HABITAT** Found in the region only in montane forest at *c.* 900–1,900m, in the canopy and mid-levels of upper and lower montane forest and along forested roadsides; occasionally down to 820m in hill forest. Surprisingly, considering that the hill stations are so well visited by birdwatchers, its nesting behaviour is unknown. The male and female both sing a wide range of beautiful notes, including imitations of other species.

Male *Female*

Asian Fairy-bluebird ▪ *Irena puella* (Murai Gajah) 25cm

DESCRIPTION In male, most of face, throat and underparts are black; largely black wings; crown, back, innerwing coverts, rump, vent and tail coverts brilliant, glossy sky-blue. Female a deep, dark powder-blue all over. Both sexes have reddish eyes and are quite bulky. **DISTRIBUTION** Resident from the central Himalayas through much of India and Sri Lanka, S China and Southeast Asia to Peninsular Malaysia, Singapore, Sumatra, Borneo and Java. **HABITS AND HABITAT** In the canopy and middle storey of lowland evergreen rainforest, peat-swamp forest and secondary woodland, less often at high altitudes into lower and even stunted upper montane forest to 1,900m. Takes many species of fruits and invertebrates, often snatching food while in flight, and seen singly or in pairs except at

major fig and fruit trees, where numbers can gather. Nesting c. Feb–Jun. Its song is much less varied and prolonged than leafbird songs.

LEFT: *Female*. ABOVE: *Male*

Baya Weaver ▪ *Ploceus philippinus* (Tempua) 15cm

DESCRIPTION Chequered brown and blackish back and wings, plain rump and russet-brown breast. In male, entire cap is deep yellow; ear coverts and throat blackish. Female has faintly striped brown crown, and buffy-brown throat, face and eyebrow.
DISTRIBUTION Resident from India and Sri Lanka to S China and parts of Southeast Asia as far as Peninsular Malaysia, Singapore, Sumatra, Java and Bali.
HABITS AND HABITAT Among tall coconut palms, grassland and the edges of secondary woodland or oil-palm estates, where tall grass is adjacent to suitable nesting trees. These trees are typically acacias, or sometimes coconut palms or bamboo, with several to many males building 1 to several nests each (a flask of drying grass, cunningly woven, suspended from a slender branch), some of which will be chosen by females for laying. In decline, through poaching of nests for sale and loss of habitat.

Java Sparrow ■ *Padda oryzivora* (Ciak Jawa) 16cm

DESCRIPTION Black crown, throat, primaries and tail; large, rounded white patch on ear coverts. Breast, back and wings light grey, sharply delineated from pink lower breast and belly. Bill and eye-ring rose-red. **DISTRIBUTION** Naturally resident only in Java and Bali, but widely introduced by man in tropical Asia and the Americas. Feral in scattered localities in Peninsular Malaysia, Singapore, Sabah and Sarawak. **HABITS**

AND HABITAT Introduced populations tend to boom for a few years after their arrival, then gradually decline, sometimes persisting for decades in very low numbers. They are able to survive only where low-intensity agriculture or careless handling of grain provides food, and old or crannied buildings provide nest sites. These conditions are now rare throughout the region, and survival is supplemented by the release of birds at religious festivals.

Dusky Munia ■ *Lonchura fuscans* (Pipit Hitam) 10cm

DESCRIPTION Small bird, dark chocolate-brown all over; blue-grey feet; dark upper and silvery-grey lower mandibles. **DISTRIBUTION** Resident and endemic to Borneo,

including 1 Philippine island (Cagayan de Sulu) that lies within Borneo coastal waters. **HABITS AND HABITAT** Singly, in pairs or small groups, never large flocks, from the coast to 1,600m, in old cultivation and forest edges. Requires more heavily vegetated habitats than other munias, and is less of a rice-eater, instead taking various grass seeds and fragments from the ground, including some apparently from animal dung. Nests of grass are built in dense vegetation or in crannies in earth banks.

Scaly-breasted Munia
■ *Lonchura punctulata* (Pipit Pinang) 10cm

DESCRIPTION Head, throat, back and wings chestnut, darkest around face; breast and belly white, strongly scalloped with black, the scallops extending to the rump. Juveniles sandy brown all over like most munias, best identified by association with adults. **DISTRIBUTION** Resident from the Himalayas through S China and Southeast Asia as far as the Lesser Sundas; introduced to various tropical areas. In Peninsular Malaysia, Singapore and, since 1993, Sabah, where it is spreading. **HABITS AND HABITAT** Associated with human cultivation, especially paddy fields, grassland, old mining land and suburban areas. Many types of grass seed are eaten, as well as ripening rice, and large flocks can build up. Nest is a grassy oval in dense vegetation; nesting has been recorded nearly throughout the year, with fewest records in the rainy season.

Black-headed Munia ■ *Lonchura atricapilla* (Pipit Rawa) 11cm

DESCRIPTION Entire head and throat black; entire body, wings and tail bright rufous chestnut. Bill grey, flushing bright turquoise during breeding. **DISTRIBUTION** Resident from India through S China to Taiwan, and throughout Southeast Asia to the Philippines, Java, Bali and the Malukus.
HABITS AND HABITAT Probably now the commonest munia, found in damp grassland, tall, scrubby grass over abandoned land, paddy fields and suburban areas, to a maximum 1,650m. Large flocks can occur in ripe rice, but pairs split off to breed at any time of year, the nest being a ball of grass among shrubby plants, or in tall grass, or even in the crown of a palm tree. At a resort hotel in Kota Kinabalu, Sabah, in 2006, every bougainvillea on the balcony of every room at each storey had 2 or 3 nests of this and the Dusky Munia (p. 108) interspersed, totalling hundreds of current or past nests.

Eurasian Tree-sparrow ■ *Passer montanus* (Ciak Rumah) 14cm

DESCRIPTION Chestnut cap and small black bib, separated by grey-white cheeks with a black spot on ear coverts. Back and wing coverts streaky brown with black streaks, and black and white wing bar; underside buffy grey. Sexes are alike, juveniles similar but duller

DISTRIBUTION Occurs discontinuously from Europe through central Asia to Japan, southwards to the Philippines and Indonesia, and throughout the region. **HABITS AND HABITAT** Commonly associated with towns and villages, factories and ports, often in large flocks, feeding on short grass and pavements, and in roadside bushes; also in rural habitats to a maximum of 1,400m within the region. After their initial colonisation, populations tend to decline again if the habitat is too well managed, especially where nesting opportunities are lost in crevices and roofs of old buildings. Eats grass seeds, spilt food and any tiny fragments picked from the ground, feeding both by day and night in brightly lit places.

Paddyfield Pipit ■ *Anthus rufulus* (Ciak Padang) 16cm

DESCRIPTION Fairly slim, upright pipit present year-round in the region, with a well-spotted breast. Very similar **Richard's Pipit** *A. richardi* is bigger and less heavily spotted,

typically has a single-note *shreep* flight call, and is probably scarce as a winter visitor. **DISTRIBUTION** India and S China through Indo-Malaya to Peninsular Malaysia, Singapore, Sumatra, Borneo, Java, the Philippines and Lesser Sundas. **HABITS AND HABITAT** On short grass such as airfields and golf courses, often with wagtails, foraging for small insects, flies, grasshoppers and spiders. Runs forward and draws itself upright on halting, often on a tussock or soil clod. Nests are well hidden among grass. Typically gives a 3-note *tchep tchep tchep* flight call when disturbed.

Grey Wagtail ▪ *Motacilla cinerea* (Kedidi Batu) 19cm

DESCRIPTION Slim, with long tail that bobs up and down; grey back, darker grey crown and sides of face, white brow; pale yellowish beneath, leading to yellow vent; white sides of tail. In flight, shows a white wing bar and yellow rump. **DISTRIBUTION** From N Africa through Europe and Asia to Japan, migrating southwards to Africa, India and Southeast Asia as far as N Australia. Migrant in Peninsular Malaysia, Singapore, Sabah and Sarawak. **HABITS AND HABITAT** Typically seen singly on the ground, on unfrequented roads, logging tracks and, especially, near streams, at any altitude from sea-level to mountains. Runs after small insects and other invertebrates on the surface of rocks, at the water's edge or even on wet tarmac, with tail wagging intermittently. A sharp 2-syllable call, *chit-chit*, is given when taking flight, unlike Eastern Yellow Wagtail (below), which usually utters a single syllable.

Eastern Yellow Wagtail ▪ *Motacilla tschutschensis* (Kedidi Kuning) 18cm

DESCRIPTION Adults at all seasons show olive-green back and rump, yellow underparts, and dark crown and face mask, but vary greatly in tone (especially head, and extent and colour of supercilium). Female duller than male, and non-breeding birds duller than breeding. Juvenile can be virtually grey and white, always with face mask meeting nape, not defined by pale rim behind.
DISTRIBUTION Resident virtually throughout N Africa, Europe and Asia; migrates to all Africa, and Southeast Asia as far as New Guinea and Australia. Migrant in Peninsular Malaysia, Singapore and Borneo. **HABITS AND HABITAT** Often seen on short grass, including coastal grassland, lawns and golf courses; forms communal roosts in wet reedbeds and at sewage farms, and by day forages singly or in small, loose groups spaced out over habitat. The 1st migrants arrive in Sep; last departures in early May. Typically makes single-note calls when taking flight.

Velvet-fronted Nuthatch

■ *Sitta frontalis* (Burung Patuk Dahi Hitam) 13cm

DESCRIPTION Bright purplish blue above and pearly grey below, with bright red bill and feet. Black velvety stubble on forehead at base of bill, continuing backwards through eye to nape in male only; red eye-ring. **DISTRIBUTION** Resident from E Himalayan foothills, discontinuously through parts of India and Sri Lanka, to S China and Southeast Asia as far as Peninsular Malaysia, Singapore, Sumatra, Borneo, Palawan and Java. **HABITS AND HABITAT** Seen in pairs or, often, small groups, including within mixed foraging flocks. Favours the upper storey of tall evergreen rainforest, from the extreme lowlands to 1,150m in lower montane forest in the Malay Peninsula, and even to 2,200m on Mount Kinabalu, Borneo (where the Blue Nuthatch, below, is absent). Forages on the larger boughs. Nesting is reported in Feb–Jun.

Blue Nuthatch

■ *Sitta azurea* (Burung Patuk Gunung) 13cm

DESCRIPTION White chin, throat and upper breast are sharply demarcated from rest of blackish head and belly; back, wings and tail increasingly blue towards rear, with bright blue edges to wing feathers. White iris and eye-ring. **DISTRIBUTION** Resident in the mountains of the far S of Peninsular Thailand, Peninsular Malaysia, Sumatra and Java. **HABITS AND HABITAT** Found in the middle and upper storeys of lower and upper montane forest at *c.* 900–1,960m, presumably displacing the Velvet-fronted Nuthatch (above) from such altitudes in the Malay Peninsula. In pairs or small parties, foraging on trunks and boughs for a variety of invertebrates. Very little is recorded about its breeding.

Javan Myna ■ *Acridotheres javanicus* (Tiong Jawa) 25cm

DESCRIPTION Ashy grey, the head darker, with yellow bill and cream iris but no bare skin around eye. White vent, wing-patch and tips to tail. **DISTRIBUTION** Formerly endemic to Java and Bali, but spread or deliberately introduced to Sumatra, Singapore and Peninsular Malaysia, scattered points in Sarawak, Sabah and Kalimantan, and the Lesser Sundas. **HABITS AND HABITAT** The bulk of Peninsular Malaysia was colonised from the 1970s onwards, by northward spread through agricultural land and along highways from Singapore. Now it is the commonest myna in the S half of the Malay Peninsula, and rapidly expanding in Sabah. Related **Makassar Myna** A. *cinereus* and possibly hybrids around Kuching and Tawau.

Common Myna ■ *Acridotheres tristis* (Tiong Rumah) 25cm

DESCRIPTION Smooth, plummy cinnamon-brown, the head nearly black, with yellow bill and bare yellow skin around dark eye. White vent, large white wing-patch, and white tips to tail. In moult, some birds have an entirely bare, scrawny yellow head. **DISTRIBUTION** Resident from the Middle East through India and S China to Southeast Asia. Invaded Peninsular Malaysia and Singapore over the past century, then Sumatra, and recently in small numbers in Sarawak. **HABITS AND HABITAT** At one time the commonest myna in Peninsular Malaysia and Singapore. Associated with human agriculture and settlements, foraging on the ground (especially on short turf) for grubs; also takes fruits and rubbish from dumps. Forms communal roosts, often with crows and other starlings and mynas; numbers have been severely reduced by invasion of Javan Myna (above).

Asian Glossy Starling ■ *Aplonis panayensis* (Perling Mata Merah) 20cm

DESCRIPTION Adults black with an oily green gloss all over, and bright red eye. Juveniles mostly blackish olive above, and heavily streaked white and blackish below, the whole plumage streaky overall and the iris dark.

DISTRIBUTION Resident from the Bay of Bengal through Myanmar and Thailand to Peninsular Malaysia and Singapore, Sumatra, Borneo, Java, Bali, the Philippines and N Sulawesi. **HABITS AND HABITAT** From mangroves and coasts, it has spread inland with human intrusion into all habitats except dryland forest, and is especially abundant in towns and villages, parkland and secondary woodland. It feeds on fruits such as overripe papayas and palm fruits, insects and nectar of the African Tulip Tree *Spathodea campanulata* (a common invasive species in the region), and forms large communal roosts, often with mynas and crows.

Common Hill-myna
■ *Gracula religiosa* (Tiong Besar) 29cm

DESCRIPTION Entirely glossy blue-black, relieved by a small white patch at base of primaries; yellow legs; rosy-yellow bill; bare yellow skin lappets on head, 1 hanging below and behind eye, another from behind eye and joining around nape. In some subspecies the lower and hind lappets join, and the situation is further confused by the release of imported cagebirds. **DISTRIBUTION** Resident from the central Himalayan foothills through S China and Southeast Asia to Peninsular Malaysia, Singapore, Sumatra, Borneo, Java, Bali to the Lesser Sundas. **HABITS AND HABITAT** In the canopy of evergreen rainforest in the extreme lowlands, extending up hill slopes only on some of the offshore islands. Even parkland is acceptable habitat if tall trees are sufficiently dense and poaching is controlled. Tremendous whistles and gurgles are made by interacting pairs, especially at fruiting fig trees.

Bornean Bristlehead
▪ *Pityriasis gymnocephala* (Tiong Batu Kepala Merah) 25cm

DESCRIPTION Smoky-black body, wings and tail, with a scalloped effect; crown with short orange stubble; rounded black patch on ear coverts; rest of head and neck brilliant carmine-red, including bare skin around eye; red thighs. Females have some red blotching on flanks. Juveniles have red ear coverts, but most of head dull blackish, and thighs black. **DISTRIBUTION** Resident and endemic in Borneo. **HABITS AND HABITAT** The chunky black bill with its fierce hooked tip enables feeding on hard, massive insects such as beetles and cicadas. Small parties overfly the canopy of peat-swamp and lowland evergreen rainforest to a maximum 1,200m; hard to detect by ground-based observers, but the whirring, croaking and nasal whines are distinctive once learned. Occasionally enters the middle and even lower storeys, peering and craning around to spot food.

Blue Rock-thrush ▪ *Monticola solitarius* (Murai Tarum) 22cm

DESCRIPTION Male plain, dark blue-grey; during breeding season, variable light and dark scale-like markings develop on body plumage. Migrant race has chestnut belly from lower breast to vent. Resident and migrant females both light uniform scaly brown all over. **DISTRIBUTION** From the Mediterranean across temperate Europe and Asia to Korea and Japan, southwards to Peninsular Malaysia and the Philippines; migrates to Africa, and S and Southeast Asia as far as the Malukus. Local resident in parts of Peninsular Malaysia; chestnut-bellied race is a scarce migrant in Peninsular Malaysia, Singapore, Sabah and Sarawak. **HABITS AND HABITAT** Residents are most often seen near limestone and other cliffs; migrants occur anywhere, but often near buildings, roadside cuttings, dams and other exposed faces, or near the seashore.

Female

Male

Chestnut-naped Forktail ■ *Enicurus ruficapillus* (Cegar Tengkuk Merah) 20cm

DESCRIPTION White forehead and rich chestnut crown; black back (brown in female), wings and throat; long, forked tail with white edges and bars; white wing bar; scaly

white breast. Females duller than males, juveniles duller still. **DISTRIBUTION** Resident from 15°N in Thailand southwards through Peninsular Malaysia, Sumatra and Borneo. **HABITS AND HABITAT** In lowland evergreen rainforest, mainly on hill slopes to *c.* 900m. An observer's 1st view is usually of a pied bird with a flash of chestnut, speeding away low over a rocky stream in the forest. Settles on rocks, especially near swirls around timber snags in the water, and often cocks, fans and lowers tail. Build nest in a rock cleft or bank, in most months of year. Call is a piercing whistle.

Slaty-backed Forktail ■ *Enicurus schistaceus* (Cegar Tengkuk Kelabu) 23cm

DESCRIPTION Adult has a black face below eye to throat, grey crown and mantle, and white rump; wings and tail black with white bars and tips to tail feathers; underside white.

Juvenile light brown above and paler, scaly brown below, with same wing and tail pattern. **DISTRIBUTION** From N India through S China and Indochina to the Malay Peninsula. Resident in Peninsular Malaysia; absent from Singapore, Sabah and Sarawak. **HABITS AND HABITAT** In hilly and montane forest and forest edges, along small rocky streams or wet roadsides in secluded areas, at *c.* 600–1,300m in our region.

Juvenile

Bornean Forktail ■ *Enicurus borneensis* (Cegar Dahi Putih Borneo) 28cm

DESCRIPTION Coal-black on face, crown, back and upper breast, with a white rump and white lower breast to vent; wings black with a single broad white cross-bar; tail long and black with white tips. On head, white is confined to a small, erect, mobile patch on forehead. The lowland **Malayan Forktail** *Enicurus frontalis* of Peninsular Malaysia and Borneo has a proportionately shorter tail, and a larger patch of white on forehead and forecrown. **DISTRIBUTION** Resident in mountains of Borneo; absent from Peninsular Malaysia and Singapore. **HABITS AND HABITAT** In Sarawak occurs only in the higher mountains, but in Sabah is found from 2,000m down to *c*. 900m, and in parts of E Sabah possibly as low as 200m. Along streams in forest and sometimes at wet roadsides or culverts.

Bornean Whistling-thrush ■ *Myophonus borneensis*
(Tiong Biru Borneo) 23cm

DESCRIPTION Male very dark slaty blue, gradually shading to brownish on wing-tips and tail; bill and legs black. Female similar but more chocolate-brown in tone. **DISTRIBUTION** Endemic resident in Borneo mountains, including Sabah and Sarawak. The **Sumatran Whistling-thrush** *M. castaneus* is closely related. **HABITS AND HABITAT** Solitary, in the middle and lower storeys of montane forest, mainly at 1,000–2,800m but extending down hill slopes to the extreme lowlands, especially where limestone caves provide nesting sites. Forages for litter-surface invertebrates and on lower parts of tree trunks, trotting forwards and then fanning the lowered tail. Call is a long-sustained monotone whistle.

Blue Whistling-thrush ■ *Myophonus caeruleus* (Tiong Belacan) 32cm

DESCRIPTION Very large thrush, black all over with a strong blue gloss; brighter blue spangles on wing coverts, and (depending on race) on back and breast. Bill bright yellow,

feet grey. The Malaysian Whistling-thrush *M. robinsoni* is a montane species and lacks any speckling. **DISTRIBUTION** From central Asia southwards through S China to Southeast Asia and Peninsular Malaysia (not S of 3°N and not in Singapore), as far as Sumatra and Java but not in Borneo. **HABITS AND HABITAT** Resident around forested limestone outcrops, where it feeds on large common snails, leaving conspicuous middens of broken shells. Nests in rock crevices, even within cave mouths. Dispersal away from limestone is limited, with records into nearby mangroves (e.g. Langkawi), along forested streams, and locally at the lowland–montane transition. Call is an intense 1–3-note whistle.

Pale Blue Jungle-flycatcher ■ *Cyornis unicolor* (Sambar Biru Muda) 17cm

DESCRIPTION Male powder-blue, paler and greyer on belly and vent. Female with rufous tail and upper-tail coverts, otherwise largely ashy grey, whiter on belly. The

similar **Verditer Flycatcher** *Eumyias thalassinus* is more turquoise (both sexes), with brighter wings and distinct black lores. **DISTRIBUTION** From NE India through S China to the Malay Peninsula, Sumatra, Borneo and Java. Resident in Peninsular Malaysia, Sabah and Sarawak; absent from Singapore. **HABITS AND HABITAT** Favours forests and the edges of forest clearings, in the middle and upper storeys, often on lower hill slopes at *c.* 200–900m, but recorded at extremes of nearly sea-level up to 1,400m. Rather inconspicuous, but shows typical flycatching behaviour from an exposed perch.

Hill Blue Flycatcher ■ *Cyornis caerulatus* (Sambar Biru Bukit) 15cm

DESCRIPTION Male deep blue above and orange-rufous below, shading gradually to pale belly and vent; no black on chin. Female brown above, slightly more rufous on wings and tail, and orange-rufous below, also shading steadily paler downwards like the male. **DISTRIBUTION** From SW China through Indochina and Malay Peninsula. Resident in Peninsular Malaysia; not Singapore. Related endemic **Javan** *C. banyumas* in Java and **Dayak** *C. montanus* in Borneo. **HABITS AND HABITAT** A bird of hill forest from *c.* 400m upwards, extending to *c.* 1,200m in lower montane forest. Flycatches in the middle and lower-middle storeys of the forest, and is usually seen singly. Its distribution seems rather patchy, and it is thought to be scarce in parts of Sabah but common in other forested areas at suitable altitudes.

Malaysian Jungle-flycatcher ■ *Cyornis turcosus* (Sambar Biru Malaysia) 14cm

DESCRIPTION Deep blue above, with a shining forehead and brow; light rufous below, shading steadily to white lower breast, belly and vent. Female is rufous all the way up to the bill; male has a black chin and throat. **DISTRIBUTION** Resident in southernmost Peninsular Thailand and Peninsular Malaysia (not Singapore), and in Sumatra and Borneo (including Sabah and Sarawak). **HABITS AND HABITAT** Found in evergreen rainforest in the extreme lowlands, in the understorey near streams and in peat-swamp forest. Claims of higher altitudes may be due to confusion with very similar species. Seen singly or in pairs, sallying out to catch passing insects. Nesting recorded *c.* Apr–Jun, with fledglings still tended by parents in Sabah until early Sep.

Female

Large Niltava ■ *Niltava grandis* (Sambar Raya) 21cm

DESCRIPTION Large flycatcher, the male very deep, dark violet, virtually black below, with brighter blue highlights on crown and sides of neck. Female rich dark brown, faintly

streaked on face and upper breast, with bluish crown and blue patch on sides of neck. **DISTRIBUTION** Resident from the central Himalayas to Yunnan, and in uplands of Southeast Asia to Peninsular Malaysia and Sumatra. **HABITS AND HABITAT** Seen singly or in pairs in lower and, less often, upper montane forest at 1,200–2,050m, often in mixed foraging flocks with other birds. Perches in the middle storey and at forest edges, sallying out to catch flying insects. Nesting is estimated to occur Feb–Jul.

Female   *Male*

Yellow-rumped Flycatcher ■ *Ficedula zanthopygia* (Sambar Belakang Kuning) 13.5cm

DESCRIPTION Male brilliant black and yellow with yellow rump; separated from male Narcissus Flycatcher *F. narcissina* by white (not yellow) brow and white wing bar extending down secondaries. Female is grey above and scaled buffy white below, with

white wing bar and yellow rump. **DISTRIBUTION** Resident in Siberia, Mongolia and China, migrating to Peninsular Malaysia, Singapore, Sumatra, Borneo and Java. **HABITS AND HABITAT** Edges of lowland evergreen forest and plantations, gardens and roadside trees, where it forages for insects on the foliage by perching, snatching or hovering. The bulk of migrants arrive from mid-Sep onwards, and most departures are in Mar–May, overflying the forest habitat. Usually solitary, foraging in the evenings.

Mugimaki Flycatcher ■ *Ficedula mugimaki* (Sambar Mugimaki) 13cm

DESCRIPTION Male blackish grey with a short white brow behind eye, and white wing-patch; rufous orange below, shading gradually paler to belly and vent. Female greyish brown above, with 2 narrow, pale wing bars; light rufous orange below, shading paler. **DISTRIBUTION** Cool temperate E Palaearctic, in Siberia and NE China, migrating southwards to Sundaland, the Philippines and Sulawesi. Migrant in Peninsular Malaysia, Singapore, Sabah and Sarawak. **HABITS AND HABITAT** Migrants have been seen from mangroves at sea-level up to more than 1,500m, in the forest and forest edges, tall secondary woodland, parks and gardens. Perches in the middle and upper storeys of forest, and flycatches for insects.

Little Pied Flycatcher ■ *Ficedula westermanni* (Sambar Hitam-putih) 11cm

DESCRIPTION Dumpy little flycatcher. Male black and grey, with very wide white eyebrow and long white wing bar. Female grey-brown above and grey-white below, with dull rufous tail and narrow wing bar. **DISTRIBUTION** Resident from the W Himalayas through S China to Peninsular Malaysia (not Singapore), Sumatra, Java, Borneo, the Philippines, Sulawesi and the Lesser Sundas. **HABITS AND HABITAT** Occurs alone or in pairs, in the crown of lower and upper montane forests and forest edges, at *c.* 1,050 - 2,030m and reaching a maximum of 3,100m on Mount Kinabalu. Gleans insects from the foliage and also sallies out to catch passing insects in flight. Nests mostly Mar–early Jun, building a cup among epiphytes or against an embankment.

Female

Male

Indigo Warbling-flycatcher
▪ *Eumyias indigo* (Sambar Sindidara) 14cm

DESCRIPTION Bright, light blue all over, nearly black around base of bill and eye, and with blue-white forehead, shading to whitish belly. Juveniles duller, with some blue colour but warm grey-brown on breast. **DISTRIBUTION** Resident in Sumatra, Java and Borneo. **HABITS AND HABITAT** Usually found perched in an upright stance along forested roadsides, in the trees or on dense vegetation along embankments, sallying out for insects as well as foliage-gleaning, and taking some berries. In lower and upper montane forest at 900–2,650m, but commonest at *c.* 1,600–1,700m. Nest is a deep cup close to the ground, beside a bank or rock face.

Asian Brown Flycatcher ▪ *Muscicapa dauurica* (Sambar Dusun) 13.5cm

DESCRIPTION Ashy grey or grey-brown, with pale lores between bill and eye, and a pale eye-ring; throat is pale with no trace of backwards extension forming a collar (unlike in Dark-sided Flycatcher M. *sibirica*); breast pale grey-brown, sometimes with faint streaks. Black feet, and yellow base to lower mandible. Several races occur as migrants from different parts of Asia, hence there is plenty of plumage variation, but the foot and bill colour, and the pale eye-ring, are good

distinguishing features. **DISTRIBUTION** Resident throughout the whole of NE Asia, and from the Himalayas through S China and Southeast Asia to Peninsular Thailand; perhaps also resident into Peninsular Malaysia, Borneo, the Philippines and the Lesser Sundas. Migrants occur throughout the region. A breeding population has been discovered in Sabah within the past 2 decades, and others may be revealed. **HABITS AND HABITAT** Found in gardens and parkland with tall scattered trees, and forest edges. Single migrants perch on bare twigs of treetops, sallying out to snap up flies, mosquitoes, moths and other insects.

Oriental Magpie-robin
■ *Copsychus saularis* (Murai Kampung) 20cm

DESCRIPTION Male has a glossy black head, breast, back, wings and white-sided tail; those in the Malay Peninsula, Singapore and Sarawak have a brilliant white belly, while Sabah males have an entirely black belly. All have a strong white wing bar. Females are more subdued grey and cream with a wing bar. **DISTRIBUTION** Resident through the Indian sub-continent, the S 3rd of China and Southeast Asia to Peninsular Malaysia, Singapore, Sumatra, Borneo, Java, Bali and most of the Philippines. **HABITS AND HABITAT** Garden bird par excellence, with a fine and varied song given by the male or both sexes from orchard trees; also occurs in plantations, secondary woodland and mangroves. Can invade edges of heavily logged forest adjacent to cultivation, and forested riverbanks. Drops to the ground for insects, worms and small vertebrates. Breeds *c.* Jan–Jun.

White-rumped Shama ■ *Copsychus malabaricus* (Murai Rimba)
Male 28cm; female 22cm

DESCRIPTION Male has a blue-glossed black head, breast, back, wings and tail; rufous belly; white rump and white edges to long tail. Female has same pattern but is duller and shorter-tailed. Juvenile similar with buff wing spots. **DISTRIBUTION** Resident from the central Himalayan foothills patchily through the Indian sub-continent and S China to Hainan, and through Peninsular Malaysia to Singapore, Sumatra, Borneo and Java. **HABITS AND HABITAT** Favours the understorey of lowland forest from sea-level upwards, rarely to 1,200m in the lower montane zone, including overgrown plantations and secondary woodland. Nesting occurs from at least Feb to Aug, the nest a cup placed in any recess. Like magpie-robins, has been persecuted by trapping for sale as a cagebird owing to its fine, varied, sustained song. The **White-crowned Shama** *C. stricklandii* of Sabah, an endemic species or race, produces distinctive powerful bursts of disyllables.

Grey-headed Canary-flycatcher ■ *Culicicapa ceylonensis*
(Sambar Kepala Kelabu) 12cm

DESCRIPTION Grey head, throat and upper breast; remainder of underparts rather bright greeny yellow; back and wings olive-green. Sexes are alike. **DISTRIBUTION** Resident

from the Himalayas through S and central China, southwards to Peninsular Malaysia (not Singapore), Sumatra, Borneo, Java, the Lesser Sundas and Flores. **HABITS AND HABITAT** Likely to be found in territorial pairs, in the middle storey from lowland evergreen rainforest in the plains to at least 1,700m in tall upper montane forest. Usually not shy, flycatching from regular perches. In Borneo, it is a brood host of the parasitic Hodgson's Hawk-cuckoo (p. 49). Call consists of 4 notes in 2 couplets, usually followed by a 5th note.

Straw-headed Bulbul
■ *Pycnonotus zeylanicus* (Barau-barau) 29cm

DESCRIPTION Largest bulbul in the region. Grey-brown above and dusky on breast, streaked whitish; white throat separated by dark moustache from straw-coloured crown and sides of face; under-tail coverts buffy yellow. Sexes are alike. **DISTRIBUTION** Resident from 12°30'N in Peninsular Thailand, through Peninsular Malaysia and Singapore to Sumatra, Nias, Borneo and Java. **HABITS AND HABITAT** Prefers forest edges along watercourses, including forest trees and bamboo, but also occurs in disturbed vegetation at any forest edge and in dense tree clumps in rural areas. Nest is a shallow cup of small twigs and fibres, in branches of small trees at forest edges or in plantations. Decimated in most places by bird trappers, because of its wonderful song, a loud, liquid duet, but experience shows it can recover well when fully protected.

Black-headed Bulbul
▪ *Microtarsus atriceps* (Merbah Siam) 18cm

DESCRIPTION Both sexes are bright olive-green, yellower on wings and tail, and with black flight feathers, including black band on tail; tip of tail bright yellow; head glossy blue-black, with contrasting pale eye. Tail pattern is easy to see in flight. Juveniles duller, with non-contrasting brown head; usually easily identified by association with adults. **DISTRIBUTION** From NE India through SW China to the Malay Peninsula, Sumatra, Java, Borneo and Palawan. Resident in Peninsular Malaysia, Singapore, Sabah and Sarawak. **HABITS AND HABITAT** Forests and forest edges, flying in small flocks from tree to tree and over low scrub in disturbed areas such as old cultivation. Takes a wide variety of small fruits, and some insects.

Black-crested Bulbul ▪ *Pycnonotus flaviventris* (Merbah Jambul Hitam) 19cm

DESCRIPTION Dull olive-green above and olive-yellow below, with entire head and throat black; iris cream, and vertical crest black. The endemic Bornean Bulbul *P. montis* is almost the same but has a yellow throat. **DISTRIBUTION** Resident from the central Himalayan foothills and Indian sub-continent through Southeast Asia, including Peninsular Malaysia (not Singapore, except for a few escapees) and Sumatra. **HABITS AND HABITAT** Although it is a lowland bird in the N part of its range, the species is largely confined to foothills and slopes in Peninsular Malaysia, in lowland evergreen forest largely at 200–1,970m, and well into upper montane forest. Occurs singly or in pairs, flying out to snap at passing insects, and taking varied small fruits and figs. Breeds roughly Jan–Jul, building a nest in dense forest-edge creepers, climbers and ferns.

Scaly-breasted Bulbul ■ *Pycnonotus squamatus* (Merbah Dada Bersisik) 15cm

DESCRIPTION Small bulbul with a scaly, dusky breast, black and white head, and olive-yellow wings; under-tail coverts yellow. Sexes are alike. DISTRIBUTION Resident in

Peninsular Thailand and Malaysia from *c.* 8°N (not Singapore), and Sumatra, Borneo (including Sabah and Sarawak) and W Java. HABITS AND HABITAT Lowland evergreen rainforest on hill slopes to *c.* 890m is the typical habitat of this delicate-looking bulbul, which feeds on small forest fruits, soft figs and, possibly, insects. It moves about in the tree crowns, rarely low down. The bird is poorly known and its breeding has never been described. Reputed to have a cheerful whistling song, a repeated pretty trill.

Stripe-throated Bulbul ■ *Pycnonotus finlaysoni* (Merbah Kunyit) 19cm

DESCRIPTION At 1st glance this is a rather an ordinary brown bulbul, but the bright yellow under-tail coverts and, in particular, heavy yellow flecking on forehead, crown,

ear coverts, throat and upper breast are very attractive. Sexes are alike. DISTRIBUTION Resident from Yunnan in S China through Myanmar, Thailand and Indochina to Peninsular Malaysia; not in Singapore, Sabah or Sarawak. HABITS AND HABITAT In the N of its range it comes down into the lowlands, but in Peninsular Malaysia it is largely confined to hill and montane forests; hence, it is found from sea-level upwards in Thailand, but mostly between 400m and 1,750m towards the centre and S of the Malay Peninsula. Feeds on fruits and insects. Nests in small bushes Feb–Aug.

Pale-faced Bulbul ■ *Pycnonotus leucops* (Merbah Meperek) 19cm

DESCRIPTION Superficially rather like the ubiquitous Yellow-vented Bulbul (below), pale below and brown above, and with yellow under-tail coverts, but the face pattern is different: it has an extensively white face without a black eye-line, and its crown has a broader brown patch rather than a narrow black top. **DISTRIBUTION** Resident from NE India discontinuously across S China to Hainan, and in Borneo including both Sabah and Sarawak. **HABITS AND HABITAT** Only in montane forest, at *c*. 1,000–3,000m, and in forest edges; seen overflying open patches, hill rice cultivation and landslides. Its food includes various small, soft fruits such as figs and wild raspberries, and possibly insects.

Sunda Yellow-vented Bulbul

■ *Pycnonotus analis* (Merbah Kapur) 20cm

DESCRIPTION Pale below (breast whiter than in Pale-faced Bulbul *P. leucops*) and brown above, with yellow under-tail coverts; black line from bill to eye, and narrow blackish strip along crown, including short erectile crest. Sexes are alike. **DISTRIBUTION** Resident from *c*. 12°30'N in Myanmar, Indochina, Peninsular Thailand and Malaysia, Singapore, Sumatra, Java, Bali, Borneo. **HABITS AND HABITAT** One of the commonest garden birds throughout the region, feeding on fruits of plants around cultivation, e.g. figs, *Lantana*, *Melastoma*, overripe papayas and many others. Also occurs in cultivation and invading forest edges along roads into the highlands. Often seen to lift both wings when landing on overhead wires. Has a cheerful, simple song.

Olive-winged Bulbul ■ *Pycnonotus plumosus* (Merbah Belukar) 20cm

DESCRIPTION Slightly larger than Yellow-vented Bulbul (p. 127) and rather plain brown above and below; wings brown but with a more olive-yellow obscure panel when

folded; ear coverts faintly streaked. Sexes are alike. **DISTRIBUTION** Resident from *c.* 12°30'N in Myanmar, Peninsular Thailand and Malaysia, Singapore (including many offshore islands), Sumatra, Borneo, Java and Palawan. **HABITS AND HABITAT** Throughout the lowlands to *c.* 500m, in forest edges and along logging tracks, but not in undisturbed forest; also in scrub, coastal vegetation and tree plantations. Seen singly, in pairs or with young, never in big flocks, feeding on small fruits such as figs in dense, tangled vegetation. Nesting occurs Jan–Jun.

Cream-vented Bulbul ■ *Pycnonotus simplex* (Merbah Mata Putih) 18cm

DESCRIPTION Slim, small-headed bulbul, brown above and paler below, lightening to cream beneath tail; eyes white in Peninsular Malaysia, red in Sabah and Sarawak (where

it is distinguished from similar Red-eyed Bulbul, p. 129, by paler underside and vent). Sexes are alike. **DISTRIBUTION** Resident from nearly 12°N in Myanmar and Peninsular Thailand to Malaysia, Singapore, Sumatra, Bunguran, Borneo and Java. **HABITS AND HABITAT** Favours the interior understorey of lowland evergreen rainforest and peat-swamp forest; commoner in disturbed areas than in totally primary vegetation. Feeds on many kinds of small, rounded fruits, and usually occurs in pairs or, sometimes, follows mixed foraging flocks. Nesting occurs mainly Feb–Apr.

Red-eyed Bulbul ■ *Pycnonotus brunneus* (Merbah Mata Merah) 19cm

DESCRIPTION Brown above and below, warmer on breast, paler on throat and under-tail coverts; iris red. Sexes are alike. **DISTRIBUTION** Resident from *c.* 12°30'N in Myanmar and Peninsular Thailand to Malaysia, Singapore (where it is scarce), Sumatra and its offshore islands, and Borneo. **HABITS AND HABITAT** One of many common bulbul species that come together to feed on abundant figs in the middle storey and canopy of forests; otherwise, it often occurs down to lower levels in forests and forest edges, from sea-level to 900m. Single birds and pairs are the normal unit, and nesting has been reported or suspected from Feb through to Aug.

Spectacled Bulbul ■ *Pycnonotus erythropthalmos* (Merbah Mata Merah Kecil) 17cm

DESCRIPTION Yet another brown bulbul, smaller and more delicate-looking than the Red-eyed (above), and slightly paler below with buff under-tail; eye is red, circled by a ring of clear yellow skin. **DISTRIBUTION** Resident from *c.* 10°N in Peninsular Thailand through Peninsular Malaysia (and formerly in Singapore), to Sumatra, Belitung and Borneo. **HABITS AND HABITAT** Somewhat less common than other brown bulbuls such as the Red-eyed and Cream-vented (p. 128), this is also a bird of the middle and lower storeys of lowland evergreen rainforest from sea-level to *c.* 800m. Like the others, it is less common within deep forest than along edges and abandoned forest roads, taking fruits of *Macaranga*, figs and other edge plants. Nesting is thought to occur *c.* Feb–May.

Red-whiskered Bulbul

■ *Pycnonotus jocosus* (Merbah Pipi Merah) 19cm

DESCRIPTION Vertical black crest, white cheeks with a red flash behind eye and red under-tail coverts are distinctive; largely white below and brown above, with a narrow moustache stripe. Sexes are alike. **DISTRIBUTION** Naturally resident in the Indian sub-continent through S China to Hainan and N Peninsular Malaysia, but heavily traded, and introduced to Penang, Ipoh, Kuala Lumpur, Singapore and, evidently, other towns. Not in Borneo. **HABITS AND HABITAT** Where successful (and not endangered by trapping) it can reach high densities, feeding on fruits from a range of garden and forest-edge plant species as well as insects, occasionally coming to the ground. Nest is a large cup of grass and creepers, hidden in dense-foliage conifers or other trees.

Buff-vented Bulbul

■ *Iole charlottae* (Merbah Riang) 20cm

DESCRIPTION Though fairly uniform grey-brown all over, is a little greyer below, with buff beneath tail. Pale iris, and slightly elongated crest feathers that often give the crown a faintly striped appearance. Sexes are alike. **DISTRIBUTION** Resident from possibly 15°N in Thailand, then discontinuously through Peninsular Thailand and Malaysia, Sumatra, the Natunas and Borneo. **HABITS AND HABITAT** Occurs in the canopy and middle levels of lowland evergreen rainforest, occasionally into the lower storey, where it sometimes participates in mixed foraging flocks, flushing insects from the foliage and pursuing them. It also takes a variety of fruits, and there is one report of a bird eating an enormous worm longer than its own body. Nesting occurs from at least May to Jul, but probably begins considerably earlier.

Hairy-backed Bulbul ◾ *Tricholestes criniger* (Merbah Bulu Tengkuk) 17cm

DESCRIPTION Look for the yellowish sides of face and around eye, contrasting with more rufous crown, brown back and slightly mottled breast, shading to yellowish buff on belly and under-tail coverts. Sexes are alike. 'Hairs' on upper back are, in fact, fine, elongated feather shafts, invisible in the field. **DISTRIBUTION**
Resident from *c.* 13°N in Peninsular Thailand and 11°30'N in Myanmar, Peninsular Malaysia (but no records from Singapore), Sumatra, the Natunas and Borneo. **HABITS AND HABITAT** Favours the middle and lower storeys of lowland evergreen rainforest, from the extreme lowlands to *c.* 900m, and peat-swamp forest, freshwater swamp forest during dry periods, and disturbed vegetation, including logged forest and regrowth. Pairs are typically seen taking insects and small fruits, but nesting is poorly described.

Yellow-bellied Bulbul ◾ *Alophoixus phaeocephalus* (Merbah Kepala Kelabu) 20cm

DESCRIPTION Quite a large bulbul, with a puffed-out white throat, blue-grey face and short, dark grey crest; olive-brown above and bright, uniform yellow below. Sexes are alike. **DISTRIBUTION** Resident in S Indochina and from *c.* 10°N in Peninsular Thailand to Peninsular Malaysia, Sumatra, the Natunas and Borneo.
HABITS AND HABITAT
Typically found in pairs, which can be noisy and are sometimes seen flying at high speed through the understorey. Prefers primary forest, from the extreme lowlands to *c.* 760m; tolerates some forest disturbance, but then occurs at lower densities. Participates in mixed foraging flocks, taking mainly insects and some fruits, and is thought to breed *c.* Feb–Jul.

Ochraceous Bulbul ■ *Alophoixus ochraceus* (Merbah Berjanggut) 20cm

DESCRIPTION One of the larger bulbuls, with a strong bill; grey-brown above and below, slightly greyer on sides of head, with throat white and often puffed out; short, erectile crest rufous or ochre-brown. Sexes are alike. **DISTRIBUTION** Resident in S Indochina and from *c*. 14°N in Peninsular Thailand, Peninsular Malaysia, Sumatra and Borneo. **HABITS AND HABITAT** In continental Asia this is a lowland bird, but in Peninsular Malaysia, Sabah and Sarawak it is confined to slopes and mountains, from *c*. 700m upwards. It occurs in pairs, often in association with mixed foraging flocks in the middle storey of montane forest. Breeding occurs in *c*. Feb–Jul. It has a range of hoarse and sweet notes, delivered in a rather random series.

Streaked Bulbul ■ *Ixos malaccensis* (Merbah Dada Berjalur) 23cm

DESCRIPTION Gives the appearance of a cool grey bird, becoming almost white on belly and under-tail coverts; sides of head, throat and upper breast with pale streaks on centre of each feather. Sexes are alike. **DISTRIBUTION** Resident from *c*. 12°30'N in Myanmar, Peninsular Malaysia (and 1 doubtful record from Singapore), Sumatra, Lingga, Bangka and Borneo. **HABITS AND HABITAT** Occurs in the forest canopy of lowland evergreen rainforest, up slopes and just reaching into montane forest at *c*. 1,100m. Figs of various species are commonly eaten, as are flying insects such as termite swarms. Because of its preference for canopy-level perches, it is often overlooked, and its nesting behaviour is virtually unknown.

Cinereous Bulbul ■ *Hemixos cinereus* (Merbah Kelabu) 21cm

DESCRIPTION Largely grey, with a puffy white throat that is often highly visible, and a contrasting darker moustache that merges upwards into grey sides of face; crown feathers erectile. Sexes are alike. **DISTRIBUTION** Resident discontinuously to Peninsular Thailand and Malaysia, Singapore, Sumatra and Borneo. Birds from the central Himalayan foothills through S China and Indochina are now separated as **Ashy Bulbul**, *H. flavala*. **HABITS AND HABITAT** Small flocks are often seen, churring to each other as they puff out their throat feathers. Lives in forests on slopes from *c.* 400m through lower and upper montane forest to *c.* 2,000m, occasionally making long-distance dispersal movements into lowlands, even as far as Singapore. Feeds on flying insects taken while perched or snatched in flight, and on fruits. Nesting may occur Feb–Jul.

Mountain Bulbul ■ *Hypsipetes mcclellandii* (Merbah Gunung) 22cm

DESCRIPTION Sturdy bulbul, common though not very well studied. Olive-green back, wings and tail; pale grey breast and bushy head, the crown feathers brown with pale central streaks. Sexes are alike. **DISTRIBUTION** Resident from the central Himalayas through S China to Hainan, and parts of Indochina and Thailand to central Peninsular Malaysia. **HABITS AND HABITAT** In Peninsular Malaysia, it is confined to montane forest of the Main and Larut ranges, Mount Tahan and a few outliers. It is a fairly common bulbul at *c.* 100–2,100m, in the forest canopy and forest edges, often among mixed foraging flocks. Various small fruits comprise the bulk of its food, along with some insects. In pairs or small groups, nesting *c.* Mar–May.

Barn Swallow ■ *Hirundo rustica* (Layang-layang Pekan) 15–20cm

DESCRIPTION Larger swallow with a black band between reddish upper breast and clean white, buff or pinkish underparts; upperparts and tail (with whitish spots) dark blue-black, the tail deeply forked with long streamers (often missing during moult). **DISTRIBUTION** Resident almost throughout the N hemisphere, in Asia southwards to N Thailand and Vietnam; non-breeding migrant to all the S continents and throughout our region. **HABITS AND HABITAT** Found almost everywhere, from lowlands to mountains, over

towns, villages, agricultural land and forest. A few individuals are present in almost every month, but predominantly in Aug–Apr. Sometimes gathers in huge flocks, especially when roosting on overhead wires in well-lit, busy small towns. Numbers have declined over the decades, but the species is still considered common. Not hard to identify by its size, clean appearance and tail shape. Feeding birds may displace smaller House Swallows (below).

Pacific Swallow ■ *Hirundo tahitica* (Layang-layang Pasifik) 14cm

DESCRIPTION Small swallow previously named Pacific Swallow with reddish forehead, throat and upper breast, this transiting directly to sullied greyish-white underparts; crown, back, wings and tail (with whitish spots) dark blue-black. Juveniles are duller, browner,

and less rufous on forehead and throat. **DISTRIBUTION** Resident from India, S China and Taiwan throughout Southeast Asia to New Guinea and the W Pacific. **HABITS AND HABITAT** Abundant throughout our region at all times of year. On the wing, appears smaller, dirtier-looking and with a short-forked tail compared with Barn Swallow (above). Often perches on wires and twigs, though not in large flocks (cf. Barn Swallow), and spends most time on the wing, flying energetically over coasts and islands, open country, mangroves and forest edges. Mud nests are built under bridges, and sometimes in roof overhangs and other buildings, mostly Feb–Jun.

Red-rumped Swallow ■ *Cecropis daurica* (Layang-layang Ekor Hitam) 16–20cm

DESCRIPTION Dark blue crown, wings and deeply forked tail; sides of face and underparts mealy white, narrowly streaked blackish. Behind face to nape, and a distinct square patch on rump, are orange-rufous in adults, buff in juveniles. **DISTRIBUTION** Resident from tropical Africa through Eurasia to the Himalayas, Japan and Korea; Asian birds migrate to Peninsular Malaysia, Singapore and Sumatra, perhaps further. **HABITS AND HABITAT** Occasional birds can be picked out from wintering Barn Swallows (p. 134) by their more deliberate flight and obvious rump colour. Has become commoner over the past 30–40 years, in habitats ranging from the landward side of mangroves at sea-level to 1,250m over forested mountains, but mostly in open coastal plains, cultivation, grassland and scrub.

Rufous-bellied Swallow ■ *Cecropis badia* (Layang-layang Batu) 18–20cm

DESCRIPTION Notably large swallow. The subspecies found in Peninsular Malaysia, *C. b. badia*, is a gorgeous brick-red colour evenly over face, breast and broad, square rump; crown, nape, back, wings and tail blackish with rich blue gloss. Juveniles are slightly duller. **DISTRIBUTION** Resident from S China through Southeast Asia to Peninsular Malaysia (not yet Singapore); also recorded once from Sumatra. Other subspecies much resembling the Red-rumped Swallow (above), with pale, streaky breasts, occur as scarce migrants in Borneo and beyond. **HABITS AND HABITAT** The brick-red Malay Peninsula birds frequent limestone outcrops but forage to many kilometres beyond these breeding sites, over forest and cultivation.

Mountain Leaftoiler ■ *Phyllergates cucullatus* (Perenjak Gunung) 12cm

DESCRIPTION Green nape, back, wings and tail; white eyebrow, throat and breast, merging to yellow belly. Tiny, with a black line through eye and a long, slim bill. Forehead

and forecrown chestnut in adults, green in juveniles. **DISTRIBUTION** Resident from E Himalayan foothills through S China to Peninsular Malaysia, Sumatra, Borneo, Java, and all the way to the Philippines and Lesser Sundas. **HABITS AND HABITAT** Like a tailorbird but does not stitch leaves together to make its nest, a grassy pouch of dead leaves in tangled vegetation at the forest edge. Found in lower and upper montane forest at 1,050–2,000m in the understorey and, more typically, at disturbed edges. Song is a roughly ascending series of 5 high notes, accompanied by buzzing from the partner.

Sunda Bush-warbler ■ *Horornis vulcanius* (Cekup Keter-keter) 13cm

DESCRIPTION Entirely dull grey-brown with a faint eyebrow, the breast greyer with impressions of speckling. Legs pinkish brown, base of lower mandible yellow. Similar

Friendly Bush-warbler *Bradypterus accentor* is darker, with a blotched throat and upper breast, rufous brow and dark grey legs. **DISTRIBUTION** Resident in mountains from Sumatra to Borneo, Java and Timor. **HABITS AND HABITAT** In middle- to high-altitude montane forest, 1,450–3,700m, from Mount Kinabalu, Sabah, to Mount Mulu and Murud, Sarawak. Mouse-like, skulking among dense, low ferns and weeds along forested roadsides and landslips. Members of a pair have a long, rising and then falling whistle; also a 5-note song, *witch-a-wee-cheee-wee*.

Chestnut-crowned Warbler ■ *Seicercus castaniceps* (Cekup Mata Putih) 10cm

DESCRIPTION Chestnut central crown-stripe and eyebrows separated by a black line, but sides of face, back and underparts light grey, shading to pale yellowish on flanks and rump; 2 pale yellowish wing bars. Sexes are alike, juveniles duller.

DISTRIBUTION Resident from central Himalayan foothills through S China and Southeast Asia to Vietnam, and in the mountains of Peninsular Malaysia and Sumatra. **HABITS AND HABITAT** In lower montane forest at *c.* 900–1,380m. On mountains where the Yellow-breasted Warbler (below) is absent (N part of the Main Range), it extends up to 1,800m; apparently the 2 species partially exclude one another. Nest is built under an overhanging bank, predominantly in Jan–Jun.

Sunda Warbler ■ *Seicercus grammiceps* (Cekup Lumut) 10cm

DESCRIPTION Bright chestnut crown and sides of face, and black lateral crown-stripes, are easy identification features; entirely bright yellow below and on rump, with green wings and tail, and 2 yellow wing bars. Sexes are alike, juveniles duller.

DISTRIBUTION Resident in mountains from Peninsular Malaysia to Sumatra, Borneo, Palawan, Flores and Timor. **HABITS AND HABITAT** Small, colourful, active warbler of upper montane forest, at 1,250–2,070m in the Malay Peninsula, and reported up to 2,450m on Mount Kinabalu. Hovers to pick insects from foliage within the forest, in the canopy down to near the understorey. Nest is purse-shaped, built in a recess on a bank in Feb–Jun.

Mountain Leaf Warbler ■ *Seicercus trivirgatus* (Cekup Daun Gunung) 11.5cm

DESCRIPTION Small warbler, olive-green above and dirty yellow below; central greenish crown-stripe, bordered by wide black lateral crown-stripes; yellow brow and black line

through eye; no wing bars. DISTRIBUTION Resident in mountains of Peninsular Malaysia, Sumatra, Borneo, Java, Bali and possibly Palawan. HABITS AND HABITAT In lower and upper montane forest at 1,300–2,160m in the Malay Peninsula, and up to 3,300m on Mount Kinabalu, Sabah. Pairs or small groups glean insects from foliage. Builds a domed nest in a recess on banks or slopes, *c.* Feb–Apr.

Hume's White-eye ■ *Zosterops auriventer* (Mata Putih Rimba) 11cm

DESCRIPTION Several white-eyes are tough to distinguish, being green above with grey flanks and a yellow throat and belly. This species has a green forehead (not yellow),

uniform with rest of crown, and the grey flanks and yellow of underside are richly coloured. DISTRIBUTION Patchily from N Thailand through the Malay Peninsula to Borneo; not in Sumatra or Java. Resident in Peninsular Malaysia, Sabah and Sarawak; absent from Singapore. HABITS AND HABITAT In hill and montane forest from a little above sea-level to at least 1,700m, replacing the similar **Oriental White-eye** *Z. palpebrosus* (with a yellow forehead), which occurs in coastal mangroves. In small, chittering flocks, frequenting the forest canopy, where it flies from tree to tree seeking small insects on the foliage.

Black-capped White-eye ■ *Zosterops atricapilla* (Mata Putih Kopiah Hitam) 10cm

DESCRIPTION Dark olive-green above and grey below; white eye-ring, emphasised by blackish surround that extends over forehead and much of face; throat and vent deep yellow, variably linked by line of yellow down centre of belly; lower mandible contrasting silver. Sexes are alike. **DISTRIBUTION** Resident in the mountains of Sumatra and Borneo. **HABITS AND HABITAT** Small flocks roam over the canopy of montane forest in Sabah and Sarawak, settling in small fruiting trees, and taking fruits of these and epiphytic mistletoes. Agile, probing among lichen and epiphytes, presumably for insects, and tearing small flowers apart to gain nectar. Its known altitude limits are *c.* 900–2,150m.

Chestnut-crested Yuhina ■ *Staphida everetti* (Burung Rimba Singgara) 12cm

DESCRIPTION Small, lively bird. White below and with white marks around eye, setting it off from foxy-chestnut ear coverts, crown and crest; upperparts grey-brown. Sexes are alike. **DISTRIBUTION** Resident and endemic in Borneo, including Sabah and Sarawak. **HABITS AND HABITAT** In the canopy and middle storey of lower and upper montane forest, usually at 900–2,800m, but also on hill slopes down to 150m in the lowlands. Flocks of 30 or more individuals move through the crowns of *Macaranga* and other roadside trees, taking small insects. They keep up a continual twittering contact call. Nest is built at ground level, in a niche on a slope or forested embankment; chicks are entirely black.

Mountain Fulvetta ■ *Alcippe peracensis* (Burung Rimba Ranting Gunung) 15cm

DESCRIPTION Rounded grey head with a neat, long, narrow black eyebrow; paler grey underparts and grey-brown back, wings and tail – virtually nondescript if eyebrow is not

noticed. **DISTRIBUTION** Hill forest in Indochina, discontinuously to mountains of Peninsular Malaysia (not Singapore). **HABITS AND HABITAT** Found in the middle and lower storeys of lower montane and upper montane forest, at 800–2,000m, including forest edges, old landslides and overgrown cultivation. Forages for invertebrates and some small fruits, and nests in Jan–May or Jun. Has a flamboyant song of 4–9 notes, varying across the scale, without which it would be noticed much less often.

Grey-throated Babbler ■ *Stachyris nigriceps* (Burung Rimba Leher Kelabu) 13cm

DESCRIPTION Dark brown back, wings and tail; more ochre-buff cheeks, breast and belly; subdued pattern of white eyebrow and white malar-patch imposed over grey crown, grey throat, and blackish around and in front of eye. **DISTRIBUTION** Resident from

the E Himalayas through S China and continental Southeast Asia to Peninsular Malaysia, including Pulau Tioman but not Singapore; also in Sumatra, Lingga, the Natunas and Borneo. **HABITS AND HABITAT** Busy groups of 4 or 5 work through the dense undergrowth, fern brakes and vegetation of forested roadsides and landslips, in lowland evergreen rainforest, and lower and upper montane forest, on slopes anywhere from just above sea-level to beyond 2,000m; never in extreme lowlands over level ground. Insectivore, breeding Jan–Jul. Continual tremulous reeling trills between flock members.

Chestnut-rumped Babbler

■ *Stachyris maculata* (Burung Rimba Rembah Besar) 18cm

DESCRIPTION Somewhat gawky appearance, with straw-yellow eye and narrow blue surround; grey head, shading to chestnut rump and tail; black throat, breaking up into strong black streaks on cream breast, and fading towards vent. **DISTRIBUTION** From c. 8°N southwards through the Malay Peninsula, Sumatra and Borneo. Resident in Peninsular Malaysia, Sabah and Sarawak; former resident in Singapore, now locally extinct.

HABITS AND HABITAT One of the more striking understorey forest babblers, seen in pairs and small family parties; cooperative breeding is probably assisted by previous young. Seeks insects among hanging clusters of dead leaves. In the group, noisy choruses with a melodious *wup wup wup…* are made by male, and harsh, scratchy notes by female, often several birds joining in.

Sunda Scimitar-babbler

■ *Pomatorhinus borneensis* (Burung Rimba Paruh Melengkung) 19cm

DESCRIPTION Impressive rufous-chestnut flanks and back; throat, breast and centre of belly pure white; wings and tail dark brown. Head black with a white eyebrow; bill long, downcurved and bright yellow with black upper base. **DISTRIBUTION** Resident in southernmost Thailand and Peninsular Malaysia from 6°N (not Singapore), through Sumatra, Bangka, and Borneo. **HABITS AND HABITAT** The middle storey and canopy of lowland evergreen rainforest, from the extreme lowlands up into lower montane forest to c. 1,350m, the upper limit varying locally. Nests down in the understorey, in a niche in an earth bank or roadside cutting. Found singly or in pairs, using the curved bill to probe for insects in bark, dead wood and epiphytes. Call is a loud, fluty *Po hoi* or *Po hoi hoi*.

Fluffy-backed Tit-babbler

■ *Macronus ptilosus* (Burung Rimba Pong-pong) 16cm

DESCRIPTION Rich rufous brown, darkening from the bright crown towards back, wings and tail; throat black. Feathers of lower back and rump are elongated and stiffened, with reduced barbs giving a hair-like appearance, but this may not be visible unless birds are displaying. Skin on lores and around eye bright blue, and 2 blue inflatable throat-patches visible when birds are calling. **DISTRIBUTION** Resident from S Thailand through

Peninsular Malaysia to Singapore (now locally extinct), Sumatra and Borneo. **HABITS AND HABITAT** In primary forest, disturbed forest and forest edges from near sea-level to a maximum 200m, including dense regrowth in the understorey and lower storey near old forest trails and the edges of old landslides and treefalls. Everywhere forages for insects in tangled thickets. Usually seen in pairs, and sometimes in family groups or mixed foraging flocks. Nesting in the Malay Peninsula occurs Dec–Aug. One member of the pair (the male?) gives a frequent *punk punk-punk-punk…* accompanied by soft churring by the other sex.

Moustached Babbler ■ *Malacopteron magnirostre* (Burung Rimba Bermisai) 17cm

DESCRIPTION Nondescript grey-brown above and off-white below, browner on tail and with dark brown crown and a dark moustache stripe; legs blue-grey. Juveniles duller, the moustache stripe less contrasting. **DISTRIBUTION** From *c.* 12°N in Myanmar through

the Malay Peninsula to Sumatra and Borneo. Resident in Peninsular Malaysia, Singapore (confirmed records only in the period 1983–94), Sabah and Sarawak. **HABITS AND HABITAT** In lowland forest from nearly sea-level to *c.* 900m, to a maximum of 1,200m in Sabah, in small parties or in pairs, and sometimes in mixed foraging flocks with other species. Gleans insects from the foliage, and may briefly chase flying insects. Song is a series of *c.* 3–6 spaced whistles, all on 1 pitch or slightly descending at the end.

Scaly-crowned Babbler
■ *Malacopteron cinereum* (Burung Rimba Tua Kecil) 16cm

DESCRIPTION Grey-brown above, and whitish below without any grey breast-streaks; secondaries, rump and tail more rufous brown. Crown blackish, with bright rufous forehead that has black feather-tips like scales; legs pink. The similar **Rufous-crowned Babbler** M. *magnum* has extensive rufous forehead without the black scales, faint grey breast-streaks and blue-grey legs. **DISTRIBUTION** Resident from 11°N in Peninsular Thailand and Malaysia (not Singapore), through

Sumatra, Bunguran and Borneo. **HABITS AND HABITAT** In the lower storey of lowland evergreen rainforest, whether primary, logged or even secondary patches, singly, in pairs or in small groups joining mixed foraging flocks. Mostly in extreme lowlands, becoming scarcer up hill slopes so as to fade out by *c.* 300–500m. Song consists of varied range of notes, a series of 1 type followed by a series of another, in different combinations.

Abbott's Wren Babbler ■ *Turdinus abbotti* (Burung Rimba Riang) 16cm

DESCRIPTION Dull rufous brown all the way from crown to tail, with more intense rufous flanks and vent, and buffy-white throat and belly. Brow and face are grey, but less distinctly so than in the very similar **Horsfield's Babbler** M. *sepiaria* in the same habitat, a species with a darker crown and obscurely streaked breast. **DISTRIBUTION** From Nepal and NE India through Indochina and the Malay Peninsula to Sumatra and Borneo. Resident in Peninsular

Malaysia, Singapore, Sabah and Sarawak. **HABITS AND HABITAT** Favours the lower storey of forest and, especially, forest edges along rivers and swampy areas, including the landward edge of mangroves and palm swamps; less common in Sabah and Sarawak than in Peninsular Malaysia. Best located by its song, an alternating *wit chee chewee; wit chee chewoo*, with 1 phrase ending in an upturn and the next in a downturn. Other, more complex songs have been described from Borneo and Sumatra.

Short-tailed Babbler ■ *Pelloreum malaccense* (Burung Rimba Ekor Pendek) 14cm

DESCRIPTION Dark brown above and on its extremely short tail, darker on crown; very distinct grey sides of face, and white throat merging to peachy-buff breast and belly. Legs

always pink. **DISTRIBUTION** Resident in Peninsular Thailand from 11°N through Peninsular Malaysia, Singapore, Sumatra, and the Anamba and Natuna islands to Borneo. **HABITS AND HABITAT** Favours ground level and the understorey of lowland evergreen rainforest, to 800–900m. Nest is a fibrous cup, built within a large, curled-up leaf on the ground, perhaps in Jan–Aug. Has a very clean appearance, hopping on the ground and giving a distinctive call: varied trilling notes, beginning softly before breaking into 5 or 6 down-slurred whistles and then another 5 or 6 deliberate steady whistles.

White-chested Babbler ■ *Pelloreum rostratum* (Burung Rimba Dada Putih) 13cm

DESCRIPTION Dark brown upperparts, a little paler on crown, with pale buff on sides of face, shading to white underparts. Pink legs; slim, straight bill, dark above and blue-grey below. **DISTRIBUTION** Resident from c. 10°30'N through Peninsular Malaysia, some offshore islands of Singapore, and in Sumatra, Belitung and Borneo. **HABITS AND HABITAT** Favours the ground and understorey within lowland evergreen rainforest over floodplains, typically near water except where a few island populations lack competing babbler species, allowing limited upslope spread. Picks small invertebrates from the litter, roots and water's edge, and breaks off to whistle from 4 (rarely, Sabah and Sarawak) to 7 undulating notes, pure and cheerful. Nest is a fine leafy cup in waterside understorey.

Streaked Wren-babbler
■ *Turdinus brevicaudatus* (Burung Rimba Hujan Gunung) 14.5cm

DESCRIPTION Pale dusky brown all over, more buff below and more grey-brown above; entire underparts lightly streaked brown; crown, back and rump lightly scaled black.

DISTRIBUTION Resident from the easternmost Himalayas and S Yunnan through Southeast Asia to the mountains of Peninsular Malaysia, including Pulau Tioman but not Singapore. **HABITS AND HABITAT** Thai populations are confined to lowland forest, but in N Peninsular Malaysia and Pulau Tioman it is found on middle slopes, and further S it occurs in lower montane forest at 750–c. 2,000m altitude. An insectivore, living on the ground alone or in pairs. Also nests on the ground Dec–Jun. Gives loud 2–4-note whistles.

Black Laughingthrush
■ *Garrulax lugubris* (Burung Rimba Hitam) 26cm

DESCRIPTION Plumage entirely unglossed black, with bare blue skin behind eye and orange-yellow bill. In closely related **Bare-headed Laughingthrush** G. *calvus* from Borneo, adults (only) have bare greeny-yellow skin on crown and sides of neck. **DISTRIBUTION** Mountains of Peninsular Malaysia and Sumatra. The Bare-headed in Borneo, including Sabah and Sarawak. **HABITS AND HABITAT** An insectivore of the middle and lower storeys of montane forest at 900–1,370m in the Malay Peninsula, but to 1,800m on Mount Kinabalu, Sabah. Found in pairs, often silent but wonderful when calling, a series of loud, frog-like gulps, followed by rich, bubbling laughter. Bornean populations have been split (as **Bare-headed Laughingthrush** M. *calva*) by some authors, but similar calls, mutual responsiveness to tape recordings, and scanty head-feathering of Peninsular Malaysian birds make the split equivocal.

Spectacled Laughingthrush
■ *Garrulax mitratus* (Burung Rimba Mata Putih) 22cm

DESCRIPTION Bright rufous-chestnut cap and face, with white ring around eye and on forehead; otherwise entirely ashy grey except for a white wing-panel and chestnut vent and thighs. Borneo birds have a buff mark only below eye (not entire ring), and slightly ochre tone to breast. Bill and legs bright orange-yellow. **DISTRIBUTION** Resident in

mountains of Peninsular Malaysia, Sumatra and Borneo, including both Sabah and Sarawak. **HABITS AND HABITAT** At *c.* 850–2,000m in lower and upper montane forest, from the canopy through the middle and lower storeys of the forest, but rarely on the ground. Also found in forest edges and secondary growth, including abandoned mountain cultivation and fern brakes. Call is a repeated 2-note whistle; breaks out into a great chorus when among flock members. Nests *c.* Feb–Jul.

Sunda Laughingthrush
■ *Leucodioptron palliatus* (Burung Rimba Lohui Puru) 25cm

DESCRIPTION Smoky blue-grey head, back, breast and belly, merging into chestnut-brown wings, rump, tail and abdomen. Sky-blue skin around eye. Sexes are alike. Resembles a giant version of the common, small, lower-storey bird, Chestnut-winged Babbler *Stachyris erythroptera*, but that species is restricted

to lowland forests, never montane. DISTRIBUTION Resident only in Sumatra and Borneo, including the mountains of both Sabah and Sarawak. HABITS AND HABITAT Small groups forage on the ground and in the lower storey up to the canopy in hill forest and lower and upper montane forest at *c.* 500–2,000m (but commonest around mid-point of this range), occasionally coming out to feed on open lawns at hill stations. A frugivore, also taking some insects. Birds keep in touch with cat-like mews and a raucous whistling chorus, 1 bird beginning with soft cooing notes and then others breaking into rattles.

Malaysian Laughingthrush ■ *Trochalopteron peninsulae*
(Burung Rimba Kepala Merah) 27cm

DESCRIPTION Rich chestnut cap, throat, breast and belly; face grey, back grey-brown.
Wing intricately patterned with chestnut greater coverts, black primary coverts, and
golden fringes to flight feathers. **DISTRIBUTION** In 2007, it was suggested that the species
was endemic to mountains of Peninsular Malaysia, and it was thus split off from closely
related forms from the Himalayas and S China through Thailand and Indochina (several
were formerly grouped together under the name Chestnut-crowned Laughingthrush
Garrulax erythrocephalus, but they have

now been split into a variety of names
and put in a new genus). **HABITS
AND HABITAT** The understorey
and middle interior of dark lower
and upper montane forest, at 1,050–
2,000m; occasionally ventures out into
old cultivation and edge vegetation.
An insectivore and partial frugivore,
nesting mainly in Jan–Apr. Usually
in pairs, seldom larger groups, with
cat-like mews and loud, jumbled,
whistling duets.

Blue-winged Siva ■ *Siva cyanouroptera* (Burung Rimba Siva) 15cm

DESCRIPTION Entirely light grey above with a white throat, underparts and sides of tail;
light violet-blue sheen on wing feathers, this often hard to see in dull, misty conditions;

distinctive pale iris. Females are
slightly duller. **DISTRIBUTION**
Resident from the central
Himalayan foothills through
S China and discontinuously
through Southeast Asia to the
mountains of Peninsular Malaysia
(not Singapore). **HABITS AND
HABITAT** Small flocks occur in
the canopy and middle storey of
lower and upper montane forest
at *c.* 1,050–1,680m, passing
through cultivation and clearings
to forage in isolated large trees.
Seeks small insects among the
foliage, as well as taking small
fruits.

Silver-eared Mesia
▪ *Leiothrix argentauris* (Burung Rimba Pipi Perak) 17cm

DESCRIPTION Very colourful with a black head; silvery-white ear coverts; yellow forehead, nape, collar and breast, and yellow in wing. Wings otherwise grey with reddish bases to flight feathers; rump and under-tail coverts reddish in male, yellow in female. **DISTRIBUTION** Resident from central Himalayan foothills through S

China, discontinuously to the mountains of Peninsular Malaysia (not Singapore) and Sumatra. **HABITS AND HABITAT** Found in the canopy of lower and upper montane forest at 900–2,000m, and down into fern brakes and scrub in old cultivation. Small, noisy parties surge through the understorey, giving a whistled 8-note song, *tee-oo-wit, tee-oo-wit, tee-oo*, and other varied notes. There are regional differences between populations in Southeast Asia.

Long-tailed Sibia ▪ *Heterophasia picaoides* (Burung Rimba Ekor Panjang) 29–32cm

DESCRIPTION Smooth, dark grey, this becoming paler on belly and under-tail coverts; shows a white flash at base of wing feathers and white tips to long, graduated tail. Iris red,

feet and slim bill black. DISTRIBUTION E Himalayan foothills to S Yunnan, and discontinuously to Peninsular Malaysia (not Singapore) and Sumatra. HABITS AND HABITAT At *c.* 1,000–2,000m in lower and upper montane forest, frequenting the crown and middle storey in small parties, including mixed foraging flocks. Groups move from tree to tree, quickly crossing open spaces at the hill stations, their long tails obvious in flight. A series of low notes is given (but no varied song; cf. Silver-eared Mesia, above) as they feed on insects and, especially, fruits. A nesting season of Feb–Jul is suggested.

Oriental Reed-warbler ▪ *Acrocephalus orientalis* (Cekup Paya Besar) 19cm

DESCRIPTION Large fawn-brown warbler, creamy whitish below. Pale eyebrow runs from bill to just behind eye, bordered below by short, dark line through eye. **DISTRIBUTION** Resident in NE and E Asia, migrating S to Southeast Asia, Peninsular Malaysia and Singapore, Sumatra, Borneo and Java, through Wallacea as far as N Australia. **HABITS AND HABITAT** Hard to see well, in reedbeds, paddy fields and tall waste grassland in the lowlands; occasionally a bird will pop up to the top of the vegetation, may call briefly, and then dive down again. In Borneo, it is recorded from 25 Sep to 24 May; in the Peninsula from 28 Aug to 26 May; and in Singapore recent stay-overs have remained into late Jun. Moult and migration have been well studied in this species.

Ashy Tailorbird ▪ *Orthotomus ruficeps* (Perenjak Kelabu) 12cm

DESCRIPTION Crown, back, wings and tail ashy grey; underparts ashy grey in male, paler in female, nearly white in juvenile. Male has extensive rufous sides of face, these more restricted in female and absent in juvenile. **DISTRIBUTION** Resident in southernmost Vietnam, and from *c.* 8°30'N in Peninsular Thailand through Malaysia and Singapore, Sumatra, Belitung, Borneo and the N coast of Java. **HABITS AND HABITAT** Mangrove forests and the landward side of mangroves everywhere, as well as peat-swamp forest in Sabah and Sarawak. Also increasingly invading inland habitats, along rivers, through plantations and orchards. In mangroves, it forages from the canopy to the mud, taking small insects from the foliage as well as flying out to snatch insects passing by. Nest is built within a 1–3-leaf stitched pouch, apparently in most months but predominantly Jan–Jun.

Dark-necked Tailorbird

■ *Orthotomus atrogularis* (Perenjak Leher Hitam) 12cm

DESCRIPTION Green back, wings and tail; entirely chestnut crown extending down to eye, with grey cheeks and ear coverts; throat wth blackish streaks, these faint in female but broad and coalescing in male; rest of underparts creamy white. Beware: Common

Tailorbird (below) can show black feather bases on throat. **DISTRIBUTION** NE India and southernmost China through Southeast Asia to Singapore, Sumatra, the Anambas and Borneo. **HABITS AND HABITAT** Favours the forest canopy from the extreme lowlands to montane forest at 1,100m, and especially the understorey along logging tracks and forest edges, riverbanks and dense, tangled secondary growth to 1,400m or more. Nest is built within a 1- or 2-leaf stitched pouch. Has a rising trill, this repeated and sometimes downturned.

Common Tailorbird ■ *Orthotomus sutorius* (Perenjak Pisang) 12cm

DESCRIPTION Dark green wings, back and tail; chestnut cap darker, duller and merging more smoothly into back than in Dark-necked Tailorbird (above); light, variable streaking

from above eye to sides of face and throat. Always with chestnut thighs. Juveniles lack the brown cap, but usually show a brownish tinge on forehead. **DISTRIBUTION** Indian sub-continent and the Himalayan foothills through S China and Southeast Asia to Peninsular Malaysia, Singapore, Bintan and Java. **HABITS AND HABITAT** The most common garden tailorbird, but also widespread in cultivation, plantations, scrub, roadsides and riverbanks. Originally confined to lowlands but has now spread to at least 1,700m with agricultural expansion. Nest is built within a 1–3-leaf stitched pouch, apparently in nearly every month. Call is a rapid, repeated *chik chik chik…* in monotonously prolonged bouts.

Rufescent Prinia ■ *Prinia rufescens* (Perenjak Belukar) 12cm

DESCRIPTION Grey head with short white brow and faint eye-ring; grey-brown back and tail with pale tips to tail feathers, and rufous tinge to wings. Underparts cream with no trace of yellow. Sexes are alike, juveniles browner than adults. **DISTRIBUTION** Resident from E Himalayan foothills and parts of E India, through S China and Southeast Asia to Peninsular Malaysia, but not reaching Singapore. **HABITS AND HABITAT** Found in rank grass and shrubs along forest edges, riverbanks and forested roadsides, from the extreme lowlands to a maximum of 1,500m in the mountains; characteristic of hillier areas and denser vegetation than the Yellow-bellied Prinia (below). Nest is like a tailorbird's, between stitched leaves in a bush. Pairs duet, 1 bird calling *chiep; chiep; chiep* while its partner gives a series of 2- and 3-note calls, *chir-chir, chir-chir-chir*.

Yellow-bellied Prinia ■ *Prinia flaviventris* (Perenjak Kuning) 14cm

DESCRIPTION Grey head and ear coverts with trace of a pale eye-ring and a short white brow (in females; obscure or absent in males); back and tail olive with white tips to tail feathers. Throat white, merging to light yellow on belly. Much variation, including feather wear and absence of yellow, causes continual confusion when using regional field guides.

DISTRIBUTION Resident from Pakistan along the Himalayan foothills through S China to Taiwan, and through Southeast Asia to Peninsular Malaysia, Singapore, Sumatra, Nias, Borneo and Java. **HABITS AND HABITAT** Widespread throughout the rural lowlands, in tall, unkempt grassland, especially wet grassland with scattered shrubs, including patches of such habitat at forest or plantation edges. Presumed to be entirely insectivorous. Males give a short, rattling song, a cat-like mew of alarm, and a sputter of wingbeats that seems to be part of their display.

Symbols

R   Breeding or known to have bred; typically but not necessarily resident all year
X   Locally extinct, formerly wild resident
M   Migrant, passage migrant, non-breeding visitor
V   Vagrant, fewer than about 5 occurrences
F   Feral
FX Locally extinct, formerly feral resident
?   Refers to uncertainty over status category (e.g. R, M or V), not over presence

Global Status according to BirdLife International and IUCN Red List 2017

LC Least Concern
NT Near Threatened
VU Vulnerable
EN Endangered
CR Critically Endangered

Taxonomy follows Eaton et al. (2016)

Common English Name	Scientific Name	Peninsular Malaysia	Singapore	Sarawak	Sabah	Global Status
Anatidae						
Wandering Whistling-duck	*Dendrocygna arcuata*	–	F	–	R	LC
Lesser Whistling-duck	*Dendrocygna javanica*	R	R	R	R	LC
White-winged Duck	*Asarcornis scutulata*	X	–	–	–	EN
Cotton Pygmy-goose	*Nettapus coromandelianus*	R	M	R	R	LC
Comb Duck	*Sarkidiornis melanotos*	V	–	–	–	LC
Gadwall	*Anas strepera*	–	V	–	–	LC
Eurasian Wigeon	*Anas penelope*	V	V	V	V	LC
Mallard	*Anas platyrhynchos*	–	–	V	V	LC
Northern Pintail	*Anas acuta*	V	V	V	V	LC
Eurasian Teal	*Anas crecca*	V	V	–	V	LC
Sunda Teal	*Anas gibberifrons*	–	–	–	R	LC
Northern Shoveler	*Spatula clypeata*	V	V	V	V	LC
Garganey	*Spatula querquedula*	M	M	V	M	LC
Tufted Duck	*Aythya fuligula*	V	V	V	V	LC
Megapodiidae						
Tabon Scrubfowl	*Megapodius cumingii*	–	–	–	R	LC
Phasianidae						
Long-billed Partridge	*Rhizothera longirostris*	R	–	R	R	NT
Dulit Partridge	*Rhizothera dulitensis*	–	–	R	R	VU
Black Partridge	*Melanoperdix niger*	R	–	R	R	VU
Blue-breasted Quail	*Synoicus chinensis*	R	R	R	R	LC
Malaysian Partridge	*Arborophila campbelli*	R	–	–	–	LC

Common English Name	Scientific Name	Peninsular Malaysia	Singapore	Sarawak	Sabah	Global Status
Bornean Partridge	*Arborophila hyperythra*	–	–	R	R	LC
Chestnut-necklaced Partridge	*Tropicoperdix charltonii*	R	–	–	R	VU
Ferruginous Partridge	*Caloperdix oculeus*	R	–	R	R?	NT
Crimson-headed Partridge	*Haematortyx sanguiniceps*	–	–	R	R	LC
Crested Partridge	*Rollulus rouloul*	R	–	R	R	NT
Red Junglefowl	*Gallus gallus*	R	R	–	F	LC
Crestless Fireback	*Lophura erythrophthalma*	R	–	R	R	VU
Malayan Crested Fireback	*Lophura ignita*	R	–	–	–	NT
Bornean Crested Fireback	*Lophura rufa*	–	–	R	R	NT
Bulwer's Pheasant	*Lophura bulweri*	–	–	R	R	VU
Mountain Peacock-pheasant	*Polyplectron inopinatum*	R	–	–	–	VU
Malaysian Peacock-pheasant	*Polyplectron malacense*	R	–	–	–	VU
Bornean Peacock-pheasant	*Polyplectron schleiermacheri*	–	–	R	R	EN
Crested Argus	*Rheinardia ocellata*	R	–	–	–	NT
Great Argus	*Argusianus argus*	R	–	R	R	NT
Green Peafowl	*Pavo muticus*	X	–	–	–	EN
Podicipedidae						
Little Grebe	*Tachybaptus ruficollis*	RM	R	–	V	LC
Columbidae						
Rock Pigeon	*Columba livia*	F	F	F	F	LC
Silvery Wood-pigeon	*Columba argentina*	–	–	X	–	CR
Red Collared Dove	*Streptopelia tranquebarica*	R	F	–	–	LC
Spotted Dove	*Spilopelia chinensis*	R	R	R	R	LC
Island Collared Dove	*Spilopelia bitorquata*	–	–	–	V	LC
Barred Cuckoo Dove	*Macropygia unchall*	R	–	–	–	LC
Ruddy Cuckoo Dove	*Macropygia emiliana*	–	–	R	R	LC
Little Cuckoo Dove	*Macropygia ruficeps*	R	–	R	R	LC
Asian Emerald Dove	*Chalcophaps indica*	R	R	R	R	LC
Zebra Dove	*Geopelia striata*	R	R	R	R	LC
Nicobar Pigeon	*Caloenas nicobarica*	R	–	–	R	NT
Cinnamon-headed Green Pigeon	*Treron fulvicollis*	R	M	R	R	NT
Little Green Pigeon	*Treron olax*	R	RM	R	R	LC
Pink-necked Green Pigeon	*Treron vernans*	R	R	R	R	LC
Orange-breasted Green Pigeon	*Treron bicincta*	R	–	–	–	LC
Thick-billed Green Pigeon	*Treron curvirostra*	R	R	R	R	LC
Large Green Pigeon	*Treron capellei*	R	–	R	R	VU
Yellow-vented Green Pigeon	*Treron seimundi*	R	–	–	–	LC
Wedge-tailed Green Pigeon	*Treron sphenura*	R	–	–	–	LC
Jambu Fruit Dove	*Ptilinopus jambu*	R	M	R	R	NT
Black-naped Fruit Dove	*Ptilinopus melanospila*	–	–	–	R	LC
Green Imperial Pigeon	*Ducula aenea*	R	M	R	R	LC
Grey Imperial Pigeon	*Ducula pickeringii*	–	–	V	R	VU
Mountain Imperial Pigeon	*Ducula badia*	R	V	R	R	LC
Pied Imperial Pigeon	*Ducula bicolor*	R	M	R	R	LC
Cuculidae						
Chestnut-winged Cuckoo	*Clamator coromandus*	R	M	M	M	LC
Pied Cuckoo	Clamator jacobinus	V	V	–	–	LC
Large Hawk-cuckoo	*Hierococcyx sparverioides*	M	M	V	V	LC

Common English Name	Scientific Name	Peninsular Malaysia	Singapore	Sarawak	Sabah	Global Status
Bock's Hawk-cuckoo	*Hierococcyx bocki*	R	–	R	R	LC
Moustached Hawk-cuckoo	*Hierococcyx vagans*	R	–	R	R	NT
Malaysian Hawk-cuckoo	*Hierococcyx fugax*	R	M	R	R	LC
Northern Hawk-cuckoo	*Hierococcyx hyperythrus*	–	–	V	V	LC
Whistling Hawk-cuckoo	*Hierococcyx nisicolor*	M	M	V	V	LC
Indian Cuckoo	*Cuculus micropterus*	RM	M	RM	RM	LC
Eurasian Cuckoo	*Cuculus canorus*	–	–	–	V	LC
Oriental Cuckoo	*Cuculus optatus*	?	V	M	M	LC
Himalayan Cuckoo	*Cuculus saturatus*	M	?	M	M	LC
Sunda Cuckoo	*Cuculus lepidus*	R	–	R	R	LC
Banded Bay Cuckoo	*Cacomantis sonneratii*	R	R	R	R	LC
Plaintive Cuckoo	*Cacomantis merulinus*	R	R	R	R	LC
Sunda Brush Cuckoo	*Cacomantis sepulcralis*	R	R	R	R	LC
Horsfield's Bronze-cuckoo	*Chrysococcyx basalis*	V	V	M	M	LC
Little Bronze-cuckoo	*Chrysococcyx minutillus*	R	R	R	R	LC
Asian Emerald Cuckoo	*Chrysococcyx maculatus*	M	V	–	–	LC
Violet Cuckoo	*Chrysococcyx xanthorhynchus*	RM	RM	R	R	LC
Square-tailed Drongo-cuckoo	*Surniculus (l.) lugubris*	R	R	R	R	LC
Fork-tailed Drongo-cuckoo	*Surniculus (l.) dicruroides*	M	–	M	M	LC
Asian Koel	*Eudynamys scolopaceus*	RM	RM	RM	RM	LC
Bornean Ground Cuckoo	*Carpococcyx radiatus*	–	–	R	R	NT
Black-bellied Malkoha	*Phaenicophaeus diardi*	R	X	R	R	NT
Chestnut-bellied Malkoha	*Phaenicophaeus sumatranus*	R	R	R	R	NT
Green-billed Malkoha	*Phaenicophaeus tristis*	R	–	–	–	LC
Raffles's Malkoha	*Phaenicophaeus chlorophaeus*	R	–	R	R	LC
Red-billed Malkoha	*Phaenicophaeus javanicus*	R	–	R	R	LC
Chestnut-breasted Malkoha	*Phaenicophaeus curvirostris*	R	–	R	R	LC
Short-toed Coucal	*Centropus rectunguis*	R	–	R	R	VU
Greater Coucal	*Centropus sinensis*	R	R	R	R	LC
Lesser Coucal	*Centropus bengalensis*	R	R	R	R	LC
Hemiprocnidae						
Grey-rumped Treeswift	*Hemiprocne longipennis*	R	R	R	R	LC
Whiskered Treeswift	*Hemiprocne comata*	R	XM	R	R	LC
Apodidae						
Giant Swiftlet	*Hydrochous gigas*	R	–	R?	R?	NT
Plume-toed Swiftlet	*Collocalia affinis*	R	R	R	R	LC
Bornean Swiftlet	*Collocalia dodgei*	–	–	–	R	–
Himalayan Swiftlet	*Aerodramus brevirostris*	M	?	–	–	LC
Mossy-nest Swiftlet	*Aerodramus salangana*	–	–	R	R	LC
Black-nest Swiftlet	*Aerodramus maximus*	R	R	R	R	LC
Edible-nest Swiftlet	*Aerodramus fuciphagus*	R	R	R	R	LC
Ameline Swiftlet	*Aerodramus amelis*	–	–	–	R	LC
Silver-rumped Needletail	*Rhaphidura leucopygialis*	R	M	R	R	LC
White-throated Needletail	*Hirundapus caudacutus*	M	M	M	M	LC
Silver-backed Needletail	*Hirundapus cochinchinensis*	M	M	–	–	LC
Brown-backed Needletail	*Hirundapus giganteus*	RM	M	M	M	LC
Asian Palm Swift	*Cypsiurus balasiensis*	R	R	R	R	LC
Fork-tailed Swift	*Apus pacificus*	M	M	M	M	LC

Common English Name	Scientific Name	Peninsular Malaysia	Singapore	Sarawak	Sabah	Global Status
House Swift	Apus nipalensis	R	R	R	R	LC
Podargidae						
Large Frogmouth	Batrachostomus auritus	R	–	R	R	NT
Dulit Frogmouth	Batrachostomus harterti	–	–	R	R	NT
Gould's Frogmouth	Batrachostomus stellatus	R	–	R	R	NT
Blyth's Frogmouth	Batrachostomus affinis	R	–	R	R	LC
Bornean Frogmouth	Batrachostomus mixtus	–	–	R	R	NT
Sunda Frogmouth	Batrachostomus cornutus	–	–	R	R	LC
Caprimulgidae						
Malaysian Eared-nightjar	Lyncornis temminckii	R	R	R	R	LC
Great Eared-nightjar	Lyncornis macrotis	R	–	–	–	LC
Grey Nightjar	Caprimulgus jotaka	M	M	M	M	LC
Large-tailed Nightjar	Caprimulgus macrurus	R	R	R	R	LC
Savanna Nightjar	Caprimulgus affinis	R	R	–	R?	LC
Bonaparte's Nightjar	Caprimulgus concretus	–	–	R	R	VU
Rallidae						
Red-legged Crake	Rallina fasciata	RM	RM	RM	RM	LC
Slaty-legged Crake	Rallina eurizonoides	M	M	–	–	LC
Buff-banded Rail	Gallirallus philippensis	–	–	R	R	LC
Slaty-breasted Rail	Lewinia striata	R	R	R	R	LC
Eastern Water Rail	Rallus indicus	–	–	V	–	LC
White-breasted Waterhen	Amaurornis phoenicurus	RM	RM	RM	RM	LC
Baillon's Crake	Zapornia pusilla	M	M	M	M	LC
Ruddy-breasted Crake	Zapornia fusca	RM	R	V	V	LC
Band-bellied Crake	Zapornia paykullii	V	–	V	–	NT
White-browed Crake	Poliolimnas cinereus	R	R	R	R	LC
Watercock	Gallicrex cinerea	RM	M	M	RM	LC
Purple Swamphen	Porphyrio porphyrio	R	R	–	R	LC
Common Moorhen	Gallinula chloropus	R	R	RM	RM	LC
Common Coot	Fulica atra	V	V	–	V	LC
Heliornithidae						
Masked Finfoot	Heliopais personatus	M	M	–	–	EN
Gruidae						
Sarus Crane	Grus antigone	X	–	–	–	VU
Burhinidae						
Beach Thick-knee	Esacus magnirostris	X?	X?	R?	R?	NT
Recurvirostridae						
Black-winged Stilt	Himantopus himantopus	RM	M	M	M	LC
Pied Stilt	Himantopus leucocephalus	–	–	V	RM	LC
Pied Avocet	Recurvirostra avosetta	–	–	V	V	LC
Haematopodidae						
Eurasian Oystercatcher	Haematopus ostralegus	V	–	V	–	LC
Dromadidae						
Crab Plover	Dromas ardeola	V	–	–	–	LC
Charadriidae						
Yellow-wattled Lapwing	Vanellus malabaricus	V	–	–	–	LC
Grey-headed Lapwing	Vanellus cinereus	M	V	V	V	LC
Red-wattled Lapwing	Vanellus indicus	R	R	–	–	LC

Common English Name	Scientific Name	Peninsular Malaysia	Singapore	Sarawak	Sabah	Global Status
Northern Lapwing	Vanellus vanellus	–	–	–	V	LC
Pacific Golden Plover	Pluvialis fulva	M	M	M	M	LC
Grey Plover	Pluvialis squatarola	M	M	M	M	LC
Common Ringed Plover	Charadrius hiaticula	V	V	–	V	LC
Long-billed Plover	Charadrius placidus	V	–	–	V	LC
Little Ringed Plover	Charadrius dubius	M	M	M	M	LC
Kentish Plover	Anarhynchus alexandrinus	M	M	M	M	LC
White-faced Plover	Anarhynchus (a.) dealbatus	M	M	V	–	DD
Malaysian Plover	Anarhynchus peronii	R	R	R	R	NT
Siberian Plover	Anarhynchus mongolus	M	M	M	M	LC
Tibetan Plover	Anarhynchus atrifrons	M	M	M	M	LC
Greater Sand-plover	Anarhynchus leschenaultii	M	M	M	M	LC
Oriental Plover	Anarhynchus veredus	V	V	V	V	LC
Rostratulidae						
Greater Painted-snipe	Rostratula benghalensis	R	V	R	R	LC
Jacanidae						
Pheasant-tailed Jacana	Hydrophasianus chirurgus	M	M	–	V	LC
Bronze-winged Jacana	Metopidius indicus	M	–	–	–	LC
Scolopacidae						
Eurasian Woodcock	Scolopax rusticola	V	V	–	V	LC
Pintail Snipe	Gallinago stenura	M	M	M	M	LC
Swinhoe's Snipe	Gallinago megala	M	M	M	M	LC
Common Snipe	Gallinago gallinago	M	M	M	M	LC
Red-necked Phalarope	Phalaropus lobatus	M	V	M	M	LC
Grey Phalarope	Phalaropus fulicarius	–	–	V	–	LC
Black-tailed Godwit	Limosa limosa	M	M	M	M	NT
Bar-tailed Godwit	Limosa lapponica	M	M	M	M	NT
Long-billed Dowitcher	Limnodromus scolopaceus	–	–	–	V	LC
Asian Dowitcher	Limnodromus semipalmatus	M	M	M	M	NT
Little Curlew	Numenius minutus	–	V	V	V	LC
Whimbrel	Numenius phaeopus	M	M	M	M	LC
Eurasian Curlew	Numenius arquata	M	M	M	M	NT
Far Eastern Curlew	Numenius madagascariensis	M	M	M	M	EN
Terek Sandpiper	Xenus cinereus	M	M	M	M	LC
Common Sandpiper	Actitis hypoleucos	M	M	M	M	LC
Green Sandpiper	Tringa ochropus	M	M	M	M	LC
Grey-tailed Tattler	Tringa brevipes	M	M	M	M	NT
Spotted Redshank	Tringa erythropus	V	M	V	V	LC
Common Greenshank	Tringa nebularia	M	M	M	M	LC
Nordmann's Greenshank	Tringa guttifer	M	M	M	V	EN
Marsh Sandpiper	Tringa stagnatilis	M	M	M	M	LC
Wood Sandpiper	Tringa glareola	M	M	M	M	LC
Common Redshank	Tringa totanus	M	M	M	M	LC
Spoon-billed Sandpiper	Eurynorhynchus pygmeus	V	V	–	–	CR
Great Knot	Calidris tenuirostris	M	M	M	M	EN
Red Knot	Calidris canutus	M	M	M	M	NT
Sanderling	Calidris alba	M	M	M	M	LC
Little Stint	Calidris minuta	M	–	–	V	LC

Common English Name	Scientific Name	Peninsular Malaysia	Singapore	Sarawak	Sabah	Global Status
Red-necked Stint	*Calidris ruficollis*	M	M	M	M	NT
Temminck's Stint	*Calidris temminckii*	M	M	M	M	LC
Long-toed Stint	*Calidris subminuta*	M	M	M	M	LC
Pectoral Sandpiper	*Calidris melanotos*	V	V	–	–	LC
Sharp-tailed Sandpiper	*Calidris acuminata*	V	V	V	M	LC
Dunlin	*Calidris alpina*	V	V	–	–	LC
Curlew Sandpiper	*Calidris ferruginea*	M	M	M	M	NT
Broad-billed Sandpiper	*Calidris falcinellus*	M	M	M	M	LC
Ruff	*Calidris pugnax*	M	M	M	M	LC
Ruddy Turnstone	*Arenaria interpres*	M	M	M	M	LC
Turnicidae						
Small Buttonquail	*Turnix sylvaticus*	R?	–	–	–	LC
Barred Buttonquail	*Turnix suscitator*	R	R	–	–	LC
Yellow-legged Buttonquail *Turnix tanki*	R?	–	–	–	LC	
Glareolidae						
Long-legged Pratincole	*Stiltia isabella*	–	–	V	V	LC
Oriental Pratincole	*Glareola maldivarum*	RM	M	M	RM	LC
Small Pratincole	*Glareola lactea*	V	V	–	–	LC
Laridae						
Brown Noddy	*Anous stolidus*	R	–	M	RM	LC
Black Noddy	*Anous minutus*	–	–	V	V	LC
Sooty Tern	*Onychoprion fuscatus*	V	–	M	RM	LC
Bridled Tern	*Onychoprion anaethetus*	R	M	R	R	LC
Aleutian Tern	*Onychoprion aleuticus*	M	M	M	–	LC
Little Tern	*Sternula albifrons*	RM	RM	RM	M	LC
Gull-billed Tern	*Gelochelidon nilotica*	M	M	M	M	LC
Caspian Tern	*Hydroprogne caspia*	M	V	V	V	LC
White-winged Tern	*Chlidonias leucopterus*	M	M	M	M	LC
Whiskered Tern	*Chlidonias hybrida*	M	M	M	M	LC
Roseate Tern	*Sterna dougallii*	R	V	M	M	LC
Black-naped Tern	*Sterna sumatrana*	R	R	R	R	LC
Common Tern	*Sterna hirundo*	M	M	M	M	LC
Lesser Crested Tern	*Thalasseus bengalensis*	M	M	M	M	LC
Great Crested Tern	*Thalasseus bergii*	RM	M	M	RM	LC
Chinese Crested Tern	*Thalasseus bernsteini*	–	–	V	–	CR
Heuglin's Gull	*Larus heuglini*	V	V	–	–	-
Laughing Gull	*Larus atricilla*	V	–	–	–	LC
Little Gull	*Larus minutus*	V	–	–	–	LC
Black-tailed Gull	*Larus crassirostris*	V	–	–	V	LC
Brown-headed Gull	*Larus brunnicephalus*	M	V	–	–	LC
Black-headed Gull	*Larus ridibundus*	M	M	M	M	LC
Slender-billed Gull	*Larus genei*	V	–	–	–	LC
Stercorariidae						
South Polar Skua	*Stercorarius maccormicki*	–	–	–	V	LC
Pomarine Skua	*Stercorarius pomarinus*	M	V	–	M	LC
Arctic Skua	*Stercorarius parasiticus*	V	V	V	V	LC
Long-tailed Skua	*Stercorarius longicaudus*	M	V	–	V	LC

Common English Name	Scientific Name	Peninsular Malaysia	Singapore	Sarawak	Sabah	Global Status
Phaethontidae						
White-tailed Tropicbird	Phaethon lepturus	–	–	–	V	LC
Procellariidae						
Streaked Shearwater	Calonectris leucomelas	M	–	M	M	NT
Short-tailed Shearwater	Ardenna tenuirostris	V	V	–	–	LC
Wedge-tailed Shearwater	Ardenna pacifica	M	–	V	V	LC
Bulwer's Petrel	Bulweria bulwerii	M	–	M	V	LC
Oceanitidae						
Wilson's Storm-petrel	Oceanites oceanicus	M	–	–	–	LC
Hydrobatidae						
Swinhoe's Storm-petrel	Oceanodroma monorhis	M	M	V	V	NT
Sulidae						
Masked Booby	Sula dactylatra	V	–	V	RV	LC
Red-footed Booby	Sula sula	V	V	V	RV	LC
Brown Booby	Sula leucogaster	RM	V	M	RM	LC
Fregatidae						
Christmas Frigatebird	Fregata andrewsi	M	V	M	M	CR
Great Frigatebird	Fregata minor	V	–	V	M	LC
Lesser Frigatebird	Fregata ariel	M	V	M	M	LC
Phalacrocoracidae						
Little Cormorant	Microcarbo niger	M	–	M	–	LC
Great Cormorant	Phalacrocorax carbo	V	–	–	M	LC
Oriental Darter	Anhinga melanogaster	XV	–	R	R	NT
Threskiornithidae						
Black-headed Ibis	Threskiornis melanocephalus	V	–	V	V	NT
White-shouldered Ibis	Pseudibis davisoni	–	–	X?	–	CR
Glossy Ibis	Plegadis falcinellus	–	V	–	V	LC
Black-faced Spoonbill	Platalea minor	–	–	V?	V?	EN
Ciconiidae						
Milky Stork	Mycteria cinerea	R	–	–	–	EN
Painted Stork	Mycteria leucocephala	F	–	–	–	NT
Asian Openbill	Anastomus oscitans	V	V	–	–	LC
Woolly-necked Stork	Ciconia episcopus	X	–	–	–	VU
Storm's Stork	Ciconia stormi	R	–	R	R	EN
Lesser Adjutant	Leptoptilos javanicus	R	M	R	R	VU
Ardeidae						
Eurasian Bittern	Botaurus stellaris	V	–	–	V	LC
Yellow Bittern	Ixobrychus sinensis	RM	RM	RM	RM	LC
Von Schrenk's Bittern	Ixobrychus eurhythmus	M	M	M	M	LC
Cinnamon Bittern	Ixobrychus cinnamomeus	RM	RM	RM	RM	LC
Black Bittern	Ixobrychus flavicollis	M	M	M	M	LC
Malaysian Night-heron	Gorsachius melanolophus	M	M	M	M	LC
Black-crowned Night-heron	Nycticorax nycticorax	R	R	R?	R	LC
Rufous Night-heron	Nycticorax caledonicus	–	–	–	R	LC
Little Heron	Butorides striata	RM	RM	RM	RM	LC
Indian Pond-heron	Ardeola grayii	M	V	–	–	LC
Chinese Pond-heron	Ardeola bacchus	M	M	M	M	LC
Javan Pond-heron	Ardeola speciosa	M	V	M	M	LC

Common English Name	Scientific Name	Peninsular Malaysia	Singapore	Sarawak	Sabah	Global Status
Cattle Egret	*Ardea ibis*	RM	FM	M	M	LC
Grey Heron	*Ardea cinerea*	R	R	M	M	LC
Great-billed Heron	*Ardea sumatrana*	R	R	R?	R	LC
Purple Heron	*Ardea purpurea*	RM	RM	RM	RM	LC
Eastern Great Egret	*Ardea modesta*	RM	M	M	RM	LC
Intermediate Egret	*Ardea intermedia*	RM	M	M	RM	LC
Little Egret	*Egretta garzetta*	RM	M	RM	RM	LC
Pacific Reef-egret	*Egretta sacra*	R	R	R	R	LC
Chinese Egret	*Egretta eulophotes*	M	M	M	M	VU
Pelecanidae						
Great White Pelican	*Pelecanus onocrotalus*	V	–	–	–	LC
Spot-billed Pelican	*Pelecanus philippensis*	V	–	–	–	NT
Pandionidae						
Osprey	*Pandion haliaetus*	M	M	M	R?M	LC
Accipitridae						
Jerdon's Baza	*Aviceda jerdoni*	R	V	R	R	LC
Black Baza	*Aviceda leuphotes*	M	M	–	–	LC
Oriental Honey-buzzard	*Pernis ruficollis*	M	M	M	M	LC
Sunda Honey-buzzard	*Pernis ptilorhyncus*	R	M	R	R	LC
Bat Hawk	*Macheiramphus alcinus*	R	M	R	R	LC
Black-winged Kite	*Elanus caeruleus*	R	R	R	R	LC
Black Kite	*Milvus migrans*	M	M	V	V	LC
Brahminy Kite	*Haliastur indus*	R	R	R	R	LC
White-bellied Fish-eagle	*Ichthyophaga leucogaster*	R	R	R	R	LC
Lesser Fish-eagle	*Ichthyophaga humilis*	R	–	R	R	NT
Grey-headed Fish-eagle	*Ichthyophaga ichthyaetus*	R	R	R	R	NT
White-rumped Vulture	*Gyps bengalensis*	X	–	–	–	CR
Slender-billed Vulture	*Gyps tenuirostris*	V	–	–	–	CR
Himalayan Griffon	*Gyps himalayensis*	V	V	–	–	NT
Cinereous Vulture	*Aegypius monachus*	V	–	–	–	NT
Red-headed Vulture	*Aegypius calvus*	X	–	–	–	CR
Short-toed Snake-eagle	*Circaetus gallicus*	M	M	–	–	LC
Crested Serpent-eagle	*Spilornis cheela*	R	R	R	R	LC
Mountain Serpent-eagle	*Spilornis kinabaluensis*	–	–	R	R	VU
Eastern Marsh Harrier	*Circus spilonotus*	M	M	M	M	LC
Western Marsh Harrier	*Circus aeruginosus*	M	V	–	–	LC
Hen Harrier	*Circus cyaneus*	V	M	V	V	LC
Pied Harrier	*Circus melanoleucos*	M	M	M	M	LC
Eurasian Sparrowhawk	*Accipiter nisus*	V	–	V	–	LC
Crested Goshawk	*Lophospiza trivirgatus*	R	RM	R	R	LC
Asian Shikra	*Tachyspiza badius*	M	–	–	–	LC
Chinese Sparrowhawk	*Tachyspiza soloensis*	M	M	M	M	LC
Japanese Sparrowhawk	*Tachyspiza gularis*	M	M	M	M	LC
Besra	*Tachyspiza virgata*	V	V	R	R	LC
Grey-faced Buzzard	*Butastur indicus*	M	M	M	M	LC
Himalayan Buzzard	*Buteo burmanicus*	V	V	–	–	LC
Common Buzzard	*Buteo buteo*	M	M	–	–	LC
Greater Spotted Eagle	*Clanga clanga*	M	M	–	–	VU

Common English Name	Scientific Name	Peninsular Malaysia	Singapore	Sarawak	Sabah	Global Status
Steppe Eagle	*Aquila nipalensis*	M	M	–	–	EN
Eastern Imperial Eagle	*Aquila heliaca*	M	V	–	–	VU
Booted Eagle	*Aquila pennata*	M	M	–	–	LC
Rufous-bellied Eagle	*Lophotriorchis kienerii*	RM	M	R	R	LC
Black Eagle	*Ictinaetus malaiensis*	R	–	R	R	LC
Blyth's Hawk-eagle	*Nisaetus alboniger*	R	M	R	R	LC
Mountain Hawk-eagle	*Nisaetus nipalensis*	R	–	–	–	LC
Changeable Hawk-eagle	*Nisaetus limnaetus*	R	R	R	R	LC
Wallace's Hawk-eagle	*Nisaetus nanus*	R	–	R	R	VU
Tytonidae						
Common Barn-owl	*Tyto alba*	R	R	F	F	LC
Eastern Grass-owl	*Tyto longimembris*	–	–	–	R	LC
Oriental Bay Owl	*Phodilus badius*	R	X	R	R	LC
Strigidae						
White-fronted Scops-owl	*Otus sagittatus*	R	–	–	–	VU
Reddish Scops-owl	*Otus rufescens*	R	–	R	R	NT
Mountain Scops-owl	*Otus spilocephalus*	R	–	R	R	LC
Rajah Scops-owl	*Otus brookii*	–	–	R	R	LC
Oriental Scops-owl	*Otus sunia*	M	M	–	–	LC
Collared Scops-owl	*Otus lempiji*	R	R	R	R	LC
Mantanani Scops-owl	*Otus mantananensis*	–	–	–	R	NT
Barred Eagle-owl	*Bubo sumatranus*	R	R	R	R	LC
Dusky Eagle-owl	*Bubo coromandus*	R	–	–	–	LC
Brown Fish-owl	*Bubo zeylonensis*	R	–	–	–	LC
Buffy Fish-owl	*Bubo ketupu*	R	R	R	R	LC
Spotted Wood-owl	*Strix seloputo*	R	R	–	–	LC
Brown Wood-owl	*Strix indranee*	R	RM	–	–	LC
Bornean Wood Owl	*Strix leptogrammica*	–	–	R	R	LC
Collared Owlet	*Glaucidium brodiei*	R	–	–	–	LC
Sunda Owlet	*Glaucidium sylvaticum*	–	–	R	R	LC
Brown Boobook	*Ninox scutulata*	R	R	R	R	LC
Northern Boobook	*Ninox japonica*	M	–	M	M	LC
Short-eared Owl	*Asio flammeus*	V	V	V	–	LC
Trogonidae						
Red-naped Trogon	*Harpactes kasumba*	R	X	R	R	NT
Diard's Trogon	*Harpactes diardii*	R	X	R	R	NT
Whitehead's Trogon	*Harpactes whiteheadi*	–	–	R	R	NT
Cinnamon-rumped Trogon	*Harpactes orrhophaeus*	R	–	R	R	NT
Scarlet-rumped Trogon	*Harpactes duvaucelii*	R	–	R	R	NT
Orange-breasted Trogon	*Harpactes oreskios*	R	–	R	R	LC
Red-headed Trogon	*Harpactes erythrocephalus*	R	–	–	–	LC
Upupidae						
Common Hoopoe	*Upupa epops*	M	–	V	V	LC
Bucerotidae						
Bushy-crested Hornbill	*Anorrhinus galeritus*	R	–	R	R	LC
Oriental Pied Hornbill	*Anthracoceros albirostris*	R	R	R	R	LC
Black Hornbill	*Anthracoceros malayanus*	R	–	R	R	NT

Common English Name	Scientific Name	Peninsular Malaysia	Singapore	Sarawak	Sabah	Global Status
Great Hornbill	*Buceros bicornis*	R	–	–	–	NT
Rhinoceros Hornbill	*Buceros rhinoceros*	R	X	R	R	NT
Helmeted Hornbill	*Rhinoplax vigil*	R	–	R	R	CR
White-crowned Hornbill	*Berenicornis comatus*	R	–	R	R	NT
Wrinkled Hornbill	*Rhabdotorrhinus corrugatus*	R	–	R	R	NT
Plain-pouched Hornbill	*Rhabdotorrhinus subruficollis*	R	–	–	–	VU
Wreathed Hornbill	*Rhyticeros undulatus*	R	–	R	R	LC
Megalaimidae						
Fire-tufted Barbet	*Psilopogon pyrolophus*	R	–	–	–	LC
Lineated Barbet	*Psilopogon lineatus*	R	F	–	–	LC
Gold-whiskered Barbet	*Psilopogon chrysopogon*	R	–	R	R	LC
Red-crowned Barbet	*Psilopogon rafflesii*	R	R	R	R	NT
Red-throated Barbet	*Psilopogon mystacophanos*	R	–	R	R	NT
Mountain Barbet	*Psilopogon monticolus*	–	–	R	R	LC
Golden-throated Barbet	*Psilopogon franklinii*	R	–	–	–	LC
Black-browed Barbet	*Psilopogon oorti*	R	–	–	–	LC
Yellow-crowned Barbet	*Psilopogon henricii*	R	–	R	R	NT
Golden-naped barbet	*Psilopogon pulcherrimus*	–	–	R	R	LC
Blue-eared Barbet	*Psilopogon australis*	R	X	R	R	LC
Bornean Barbet	*Psilopogon eximius*	–	–	R	R	LC
Coppersmith Barbet	*Psilopogon haemacephalus*	R	R	–	–	LC
Bornean Brown Barbet	*Calorhamphus fuliginosus*	–	–	R	R	LC
Malayan Brown Barbet	*Calorhamphus hayii*	R	–	–	–	NT
Indicatoridae						
Sunda Honeyguide	*Indicator archipelagicus*	R	–	R	R	NT
Picidae						
Eurasian Wryneck	*Jynx torquilla*	V	–	–	–	LC
Speckled Piculet	*Picumnus innominatus*	R	–	–	R	LC
Rufous Piculet	*Sasia abnormis*	R	–	R	R	LC
Sunda Pygmy Woodpecker	*Picoides moluccensis*	R	R	R	R	LC
Grey-capped Pygmy Woodpecker	*Picoides canicapillus*	R	X	R	R	LC
Rufous Woodpecker	*Micropternus brachyurus*	R	R	R	R	LC
White-bellied Woodpecker	*Dryocopus javensis*	R	X	R	R	LC
Great Slaty Woodpecker	*Dryocopus pulverulentus*	R	–	R	R	VU
Banded Woodpecker	*Chrysophlegma mineaceum*	R	R	R	R	LC
Greater Yellownape	*Chrysophlegma flavinucha*	R	–	–	–	LC
Checquer-throated Woodpecker	*Chrysophlegma mentale*	R	X	R	R	NT
Lesser Yellownape	*Picus chlorolophus*	R	–	–	–	LC
Crimson-winged Woodpecker	*Picus puniceus*	R	X	R	R	LC
Grey-headed Woodpecker	*Picus canus*	R	–	–	–	LC
Streak-breasted Woodpecker	*Picus viridanus*	R	–	–	–	LC
Laced Woodpecker	*Picus vittatus*	R	R	–	–	LC
Olive-backed Woodpecker	*Dinopium rafflesii*	R	X	R	R	NT
Common Flameback	*Dinopium javanense*	R	R	R	R	LC
Greater Flameback	*Chrysocolaptes guttacristatus*	R	X	R	R	LC
Orange-backed Wodpecker	*Chrysocolaptes validus*	R	X	R	R	LC
Bamboo Woodpecker	*Gecinulus viridis*	R	–	–	–	LC

Common English Name	Scientific Name	Peninsular Malaysia	Singapore	Sarawak	Sabah	Global Status
Maroon Woodpecker	*Blythipicus rubiginosus*	R	–	R	R	LC
Bay Woodpecker	*Blythipicus pyrrhotis*	R	–	–	–	LC
Buff-rumped Woodpecker	*Meiglyptes tristis*	R	V	R	R	LC
Buff necked Woodpecker	*Meiglyptes tukki*	R	X	R	R	NT
Grey-and-buff Woodpecker	*Hemicircus concretus*	R	X	R	R	LC
Alcedinidae						
Rufous-collared Kingfisher	*Actenoides concretus*	R	X	R	R	NT
Banded Kingfisher	*Lacedo pulchella*	R	–	R	R	LC
Stork-billed Kingfisher	*Pelargopsis capensis*	R	R	R	R	LC
Brown-winged Kingfisher	*Pelargopsis amauropotera*	R	–	–	–	NT
Ruddy Kingfisher	*Halcyon coromanda*	RM	RM	RM	RM	LC
White-throated Kingfisher	*Halcyon smyrnensis*	R	R	–	–	LC
Black-capped Kingfisher	*Halcyon pileata*	M	M	M	M	LC
Collared Kingfisher	*Todirhamphus chloris*	RM	R	RM	RM	LC
Sacred Kingfisher	*Todirhamphus sanctus*	–	–	M	M	LC
Black-backed Kingfisher	*Ceyx erithaca*	M	M	M	M	LC
Rufous-backed Kingfisher	*Ceyx rufidorsa*	R	X	R	R	LC
Malay Blue-banded Kingfisher	*Alcedo peninsulae*	R	–	R	R	VU
Blue-eared Kingfisher	*Alcedo meninting*	R	R	R	R	LC
Common Kingfisher	*Alcedo atthis*	M	M	M	M	LC
Pied Kingfisher	*Ceryle rudis*	V	–	–	–	LC
Meropidae						
Red-bearded Bee-eater	*Nyctyornis amictus*	R	–	R	R	LC
Blue-tailed Bee-eater	*Merops philippinus*	RM	M	RM	RM	LC
Blue-throated Bee-eater	*Merops viridis*	RM	R	RM	RM	LC
Chestnut-headed Bee-eater	*Merops leschenaulti*	R	–	–	–	LC
Rainbow Bee-eater	*Merops ornatus*	–	–	–	V	LC
Coraciidae						
Indian Roller	*Coracias benghalensis*	R	–	–	–	LC
Dollarbird	*Eurystomus orientalis*	RM	RM	RM	RM	LC
Falconidae						
Black-thighed Falconet	*Microhierax fringillarius*	R	M	R	R	LC
White-fronted Falconet	*Microhierax latifrons*	–	–	–	R	NT
Lesser Kestrel	*Falco naumanni*	–	V	–	–	VU
Eurasian Kestrel	*Falco tinnunculus*	M	M	M	M	LC
Amur Falcon	*Falco amurensis*	V	V	–	–	LC
Eurasian Hobby	*Falco subbuteo*	V	V	–	V	LC
Oriental Hobby	*Falco severus*	V?	V	–	V	LC
Peregrine Falcon	*Falco peregrinus*	RM	M	RM	RM	LC
Cacatuidae						
Yellow-crested Cockatoo	*Cacatua sulphurea*	–	F	–	–	LC
Tanimbar Cockatoo	*Cacatua goffiniana*	–	F	–	–	NT
Psittacidae						
Blue-crowned Hanging Parrot	*Loriculus galgulus*	R	R	R	R	LC
Blue-rumped Parrot	*Psittinus cyanurus*	R	R	R	R	NT
Blue-naped Parrot	*Tanygnathus lucionensis*	–	–	FX	R	NT
Rose-ringed Parakeet	*Psittacula krameri*	–	F	–	–	LC
Red-breasted Parakeet	*Psittacula alexandri*	–	F	–	–	LC

Common English Name	Scientific Name	Peninsular Malaysia	Singapore	Sarawak	Sabah	Global Status
Long-tailed Parakeet	*Psittacula longicauda*	R	R	R	R	NT
Rainbow Lorikeet	*Trichoglossus haematodus*	–	F	–	–	LC
Calyptomenidae						
Green Broadbill	*Calyptomena viridis*	R	X	R	R	NT
Hose's Broadbill	*Calyptomena hosei*	–	–	R	R	NT
Whitehead's Broadbill	*Calyptomena whiteheadi*	–	–	R	R	LC
Eurylaimidae						
Long-tailed Broadbill	*Psarisomus dalhousiae*	R	–	R	R	LC
Dusky Broadbill	*Corydon sumatranus*	R	X	R	R	LC
Silver-breasted Broadbill	*Serilophus lunatus*	R	–	–	–	LC
Black-and-red Broadbill	*Cymbirhynchus macrorhynchos*	R	XM	R	R	LC
Banded Broadbill	*Eurylaimus javanicus*	R	X	R	R	LC
Black-and-yellow Broadbill	*Eurylaimus ochromalus*	R	X	R	R	NT
Pittidae						
Giant Pitta	*Hydrornis caeruleus*	R	–	R	R	NT
Malayan Banded Pitta	*Hydrornis irena*	R	–	–	–	NT
Bornean Banded Pitta	*Hydrornis schwaneri*	–	–	R	R	NT
Blue-headed Pitta	*Hydrornis baudii*	–	–	R	R	VU
Blue-banded Pitta	*Erythropitta arquata*	–	–	R	R	LC
Garnet Pitta	*Erythropitta granatina*	R	X	R	–	NT
Black-headed Pitta	*Erythropitta ussheri*	–	–	–	R	NT
Rusty-naped Pitta	*Pitta oatesi*	R	–	–	–	LC
Hooded Pitta	*Pitta sordida*	RM	M	RM	RM	LC
Fairy Pitta	*Pitta nympha*	–	–	M	M	VU
Blue-winged Pitta	*Pitta moluccensis*	RM	M	M	M	LC
Mangrove Pitta	*Pitta megarhyncha*	R	R	R?	–	NT
Pardalotidae						
Golden-bellied Gerygone	*Gerygone sulphurea*	R	R	R	R	LC
Vireonidae						
Blyth's Shrike-vireo	*Pteruthius aeralatus*	R	–	R	R	LC
Black-eared Shrike-vireo	*Pteruthius melanotis*	R	–	–	–	LC
White-bellied Erpornis	*Erpornis zantholeuca*	R	–	R	R	LC
Oriolidae						
Dark-throated Oriole	*Oriolus xanthonotus*	R	X	R	R	NT
Eurasian Golden Oriole	*Oriolus oriolus*	V	–	–	–	LC
Black-naped Oriole	*Oriolus chinensis*	RM	RM	R?M?	R?M?	LC
Black-hooded Oriole	*Oriolus xanthornus*	R	–	–	–	LC
Black-and-crimson Oriole	*Oriolus cruentus*	R	–	R	R	LC
Black Oriole	*Oriolus hosii*	–	–	R	?	NT
Pachycephalidae						
Mangrove Whistler	*Pachycephala cinerea*	R	R	R	R	LC
White-vented Whistler	*Pachycephala homeyeri*	–	–	–	R	LC
Bornean Whistler	*Pachycephala hypoxantha*	–	–	R	R	LC
Vangidae						
Bar-winged Flycatcher-shrike	*Hemipus picatus*	R	–	R	R	LC
Black-winged Flycatcher-shrike	*Hemipus hirundinaceus*	R	V	R	R	LC
Large Woodshrike	*Tephrodornis gularis*	R	X	R	R	LC

Common English Name	Scientific Name	Peninsular Malaysia	Singapore	Sarawak	Sabah	Global Status
Rufous-winged Philentoma	*Philentoma pyrhoptera*	R	X	R	R	LC
Maroon-breasted Philentoma	*Philentoma velata*	R	X	R	R	NT
Pityriaseidae						
Bornean Bristlehead	*Pityriasis gymnocephala*	–	–	R	R	NT
Aegithinidae						
Common Iora	*Aegithina tiphia*	R	R	R	R	LC
Green Iora	*Aegithina viridissima*	R	X	R	R	NT
Great Iora	*Aegithina lafresnayei*	R	–	–	–	LC
Artamidae						
Ashy Woodswallow	*Artamus fuscus*	V	–	–	–	LC
White-breasted Woodswallow	*Artamus leucorynchus*	R	–	R	R	LC
Campephagidae						
Malaysian Cuckooshrike	*Coracina larutensis*	R	–	–	–	LC
Sunda Cuckooshrike	*Coracina larvata*	–	–	R	R	LC
Roving Cuckooshrike	*Coracina sumatrensis*	R	X	R	R	LC
Lesser Cicadabird	*Lalage fimbriata*	R	R	R	R	LC
Pied Triller	*Lalage nigra*	R	R	R	R	LC
Rosy Minivet	*Pericrocotus roseus*	V	–	–	–	LC
Ashy Minivet	*Pericrocotus divaricatus*	M	M	M	M	LC
Swinhoe's Minivet	*Pericrocotus cantonensis*	V	–	–	–	LC
Fiery Minivet	*Pericrocotus igneus*	R	X	R	R	NT
Grey-chinned Minivet	*Pericrocotus solaris*	R	–	R	R	LC
Scarlet Minivet	*Pericrocotus flammeus*	R	R	R	R	LC
Rhipiduridae						
White-throated Fantail	*Rhipidura albicollis*	R	–	R	R	LC
Sunda Pied Fantail	*Rhipidura javanica*	R	R	R	R	LC
Spotted Fantail	*Rhipidura perlata*	R	–	R	R	LC
Dicruridae						
Black Drongo	*Dicrurus macrocercus*	M	M	–	V	LC
Ashy Drongo	*Dicrurus leucophaeus*	RM	M	R	R	LC
Crow-billed Drongo	*Dicrurus annectans*	M	M	M	M	LC
Bronzed Drongo	*Dicrurus aeneus*	R	X	R	R	LC
Lesser Racquet-tailed Drongo	*Dicrurus remifer*	R	–	–	–	LC
Greater Racquet-tailed Drongo	*Dicrurus paradiseus*	R	R	R	R	LC
Bornean Spangled Drongo	*Dicrurus borneensis*	–	–	R	R	LC
Monarchidae						
Black-naped Monarch	*Hypothymis azurea*	R	R	R	R	LC
Blyth's Paradise-flycatcher	*Terpsiphone affinis*	R	R	R	R	LC
Amur Paradise-flycatcher	*Terpsiphone incei*	M	M	–	–	LC
Japanese Paradise-flycatcher	*Terpsiphone atrocaudata*	M	M	–	V	NT
Laniidae						
Jay Shrike	*Platylophus galericulatus*	R	–	R	R	NT
Tiger Shrike	*Lanius tigrinus*	M	M	M	M	LC
Brown Shrike	*Lanius cristatus*	M	M	M	M	LC
Long-tailed Shrike	*Lanius schach*	R	R	M	M	LC
Corvidae						
House Crow	*Corvus splendens*	F ?	F?	–	F	LC
Sunda Crow	*Corvus enca*	R	–	R	R	LC

Common English Name	Scientific Name	Peninsular Malaysia	Singapore	Sarawak	Sabah	Global Status
Southern Jungle Crow	*Corvus macrorhynchos*	R	R	R?	R?	LC
Common Green Magpie	*Cissa chinensis*	R	–	R	R	LC
Bornean Green Magpie	*Cissa jefferyi*	–	–	R	R	LC
Bornean Treepie	*Dendrocitta cinerascens*	–	–	R	R	LC
Racquet-tailed Treepie	*Crypsirina temia*	R	–	–	–	LC
Malayan Black Magpie	*Platysmurus leucopterus*	R	–	–	–	NT
Bornean Black Magpie	*Platysmurus aterrimus*	–	–	R	R	LC
Eupetidae						
Rail-babbler	*Eupetes macrocerus*	R	–	R	R	NT
Stenostiridae						
Grey-headed Canary-flycatcher	*Culicicapa ceylonensis*	R	–	R	R	LC
Paridae						
Cinereous Tit	*Parus cinereus*	R	–	R	R	
Sultan Tit	*Melanochlora sultanea*	R	–	–	–	LC
Alaudidae						
Eurasian Skylark	*Alauda arvensis*	–	–	V	V	LC
Oriental Skylark	*Alauda gulgula*	V	–	–	–	LC
Hirundinidae						
Asian House-martin	*Delichon dasypus*	M	M	M	M	LC
Common Sand-martin	*Riparia riparia*	M	M	M	M	LC
Dusky Crag-martin	*Ptyonoprogne concolor*	R	–	–	–	LC
Barn Swallow	*Hirundo rustica*	M	M	M	M	LC
Pacific Swallow	*Hirundo tahitica*	R	R	R	R	LC
Daurian Swallow	*Cecropis daurica*	M	M	M	M	LC
Rufous-bellied Swallow	*Cecropis badia*	R	–	–	–	LC
Pycnonotidae						
Straw-headed Bulbul	*Pycnonotus zeylanicus*	R	R	R	R	EN
Black-crested Bulbul	*Pycnonotus flaviventris*	R	F	–	–	LC
Bornean Bulbul	*Pycnonotus montis*	–	–	R	R	LC
Scaly-breasted Bulbul	*Pycnonotus squamatus*	R	–	R	R	NT
Grey-bellied Bulbul	*Pycnonotus cyaniventris*	R	X	R	R	NT
Stripe-throated Bulbul	*Pycnonotus finlaysoni*	R	–	–	–	LC
Pale-faced Bulbul	*Pycnonotus leucops*	–	–	R	R	LC
Sunda Yellow-vented Bulbul	*Pycnonotus analis*	R	R	R	R	LC
Olive-winged Bulbul	*Pycnonotus plumosus*	R	R	R	R	LC
Streak-eared Bulbul	*Pycnonotus blanfordi*	R	–	–	–	LC
Cream-vented Bulbul	*Pycnonotus simplex*	R	R	R	R	LC
Asian Red-eyed bulbul	*Pycnonotus brunneus*	R	R	R	R	LC
Spectacled Bulbul	*Pycnonotus erythropthalmos*	R	–	R	R	LC
Red-whiskered Bulbul	*Pycnonotus jocosus*	F ?	F	–	–	LC
Sooty-headed Bulbul	*Pycnonotus aurigaster*	–	F	–	–	LC
Puff-backed Bulbul	*Microtarsus eutilotus*	R	–	R	R	NT
Black-and-white Bulbul	*Microtarsus melanoleucos*	R	V	R	R	NT
Black-headed Bulbul	*Microtarsus atriceps*	R	R	R	R	LC
Hook-billed Bulbul	*Setornis criniger*	–	–	R	R	VU
Buff-vented Bulbul	*Iole crypta*	R	R	R	R	NT
Hairy-backed Bulbul	*Tricholestes criniger*	R	–	R	R	LC
Finsch's Bulbul	*Alophoixus finschii*	R	–	R	R	NT

Common English Name	Scientific Name	Peninsular Malaysia	Singapore	Sarawak	Sabah	Global Status
Yellow-bellied Bulbul	*Alophoixus phaeocephalus*	R	X	R	R	LC
Grey-cheeked Bulbul	*Alophoixus tephrogenys*	R	–	R	R	LC
Ochraceous Bulbul	*Alophoixus ochraceus*	R	–	–	–	LC
Chestnut-vented Bulbul	*Alophoixus ruficrissus*	–	–	R	R	LC
Streaked Bulbul	*Ixos malaccensis*	R	M	R	R	NT
Mountain Bulbul	*Ixos mcclellandii*	R	–	–	–	LC
Cinereous Bulbul	*Hemixos cinereus*	R	M	R	R	LC
Timaliidae						
Striped Tit-babbler	*Mixornis gularis*	R	R	R	R	LC
Fluffy-backed Tit-babbler	*Macronus ptilosus*	R	–	R	R	NT
Golden Babbler	*Cyanoderma chrysaeum*	R	–	–	–	LC
Rufous-fronted Babbler	*Cyanoderma rufifrons*	R	–	R	R	LC
Chestnut-winged Babbler	*Cyanoderma erythropterum*	R	R	R	R	LC
Large Scimitar-babbler	*Pomatorhinus hypoleucos*	R	–	–	–	LC
Sunda Scimitar-babbler	*Pomatorhinus bornensis*	R	–	R	R	LC
Black-throated Babbler	*Stachyris nigricollis*	R	–	R	R	NT
White-necked Babbler	*Stachyris leucotis*	R	–	R	R	NT
Grey-headed Babbler	*Stachyris poliocephala*	R	–	R	R	LC
Grey-throated Babbler	*Stachyris nigriceps*	R	–	R	R	LC
Chestnut-rumped Babbler	*Stachyris maculata*	R	–	R	R	NT
Pellorneidae						
Moustached Babbler	*Malacopteron magnirostre*	R	R	R	R	LC
Sooty-capped Babbler	*Malacopteron affine*	R	–	R	R	NT
Scaly-crowned Babbler	*Malacopteron cinereum*	R	–	R	R	LC
Rufous-crowned Babbler	*Malacopteron magnum*	R	–	R	R	NT
Grey-breasted Babbler	*Malacopteron albogularis*	R	–	R	R	NT
Buff-breasted Babbler	*Pellorneum tickelli*	R	–	–	–	LC
Black-capped Babbler	*Pellorneum capistratum*	R	X	R	R	LC
Puff-throated Babbler	*Pellorneum ruficeps*	R	–	–	–	LC
Short-tailed Babbler	*Pellorneum malaccensis*	R	R	R	R	NT
Temminck's Babbler	*Pellorneum pyrrogenys*	–	–	R	R	LC
White-chested Babbler	*Pellorneum rostratum*	R	R	R	R	NT
Ferruginous Babbler	*Pellorneum bicolor*	R	–	R	R	LC
Striped Wren-babbler	*Kenopia striata*	R	–	R	R	NT
Abbott's Wren-babbler	*Turdinus abbotti*	R	R	R	R	LC
Horsfield's Wren-babbler	*Turdinus sepiarius*	R	–	R	R	LC
Marbled Wren-babbler	*Turdinus marmoratus*	R	–	–	–	LC
Large Wren-babbler	*Turdinus macrodactylus*	R	–	–	–	NT
Streaked Wren-babbler	*Turdinus brevicaudatus*	R	–	–	–	LC
Black-throated Wren-babbler	*Turdinus atrigularis*	–	–	R	R	NT
Mountain Wren-babbler	*Turdinus crassus*	–	–	R	R	LC
Eyebrowed Wren-babbler	*Napothera epilepidota*	R	–	R	R	LC
Bornean Wren-babbler	*Ptilocichla leucogrammica*	–	–	R	R	VU
Leiothrichidae						
Mountain Fulvetta	*Alcippe peracensis*	R	–	–	–	LC
Sunda Fulvetta	*Alcippe brunneicauda*	R	–	R	R	NT
Collared Babbler	*Gampsorhynchus torquatus*	R	–	–	–	LC
Rufous-winged Fulvetta	*Pseudominla castaneceps*	R	–	–	–	LC

Common English Name	Scientific Name	Peninsular Malaysia	Singapore	Sarawak	Sabah	Global Status
Himalayan Cutia	*Cutia nipalensis*	R	–	–	–	LC
White-crested Laughingthrush	*Leucodioptron leucolophum*	F	F	–	–	LC
Sunda Laughingthrush	*Leucodioptron palliatum*	–	–	R	R	LC
Chinese Hwamei	*Leucodioptron canorus*	–	F	–	–	LC
Black Laughingthrush	*Garrulax lugubris*	R	–	–	–	LC
Bare-headed Laughingthrush	*Garrulax calvus*	–	–	R	R	LC
Spectacled Laughingthrush	*Garrulax mitratus*	R	–	R	R	LC
Chestnut-hooded Laughingthrush	*Garrulax treacheri*	–	–	R	R	LC
Malaysian Laughingthrush	*Trochalopteron peninsulae*	R	–	–	–	LC
Bar-throated Minla	*Chrysominla strigula*	R	–	–	–	LC
Blue-winged Siva	*Siva cyanouroptera*	R	–	–	–	LC
Silver-eared Mesia	*Leiothrix argentauris*	R	–	–	–	LC
Long-tailed Sibia	*Heterophasia picaoides*	R	–	–	–	LC
Zosteropidae						
Mountain Blackeye	*Zosterops emiliae*	–	–	R	R	LC
Sunda White-eye	*Zosterops melanurus*	R	R	R	R	LC
Black-capped White-eye	*Zosterops atricapilla*	–	–	R	R	LC
Hume's White-eye	*Zosterops auriventer*	R	–	R	R	LC
Pygmy White-eye	*Oculocincla squamiceps*	–	–	R	R	LC
Chestnut-crested Yuhina	*Staphida everetti*	–	–	R	R	LC
Phylloscopidae						
JapaneseLeaf-warbler	*Phylloscopus xanthodryas*	M	–	M	M	LC
Pale-legged Leaf-warbler	*Phylloscopus tenellipes*	M?	–	–	–	LC
Sakhalin Leaf-warbler	*Phylloscopus borealoides*	–	V	–	–	LC
Two-barred Warbler	*Phylloscopus plumbeitarsus*	V	–	–	–	LC
Willow Warbler	*Phylloscopus trochilus*	–	–	–	V	LC
Yellow-browed Warbler	*Phylloscopus inornatus*	M	M	V	–	LC
Dusky Warbler	*Phylloscopus fuscatus*	M	V	–	–	LC
Radde's Warbler	*Phylloscopus schwarzi*	V	–	–	–	LC
Plain-tailed Warbler	*Seicercus soror*	V	–	–	–	LC
Chestnut-crowned Warbler	*Seicercus castaniceps*	R	–	–	–	LC
Sunda Warbler	*Seicercus grammiceps*	R	–	R	R	LC
Eastern Crowned Warbler	*Seicercus coronatus*	M	M	–	–	LC
Mountain Leaf-warbler	*Seicercus trivirgatus*	R	–	R	R	LC
Arctic Warbler	*Seicercus borealis*	M	M	M	M	LC
Cettiidae						
Bamboo Bush Warbler	*Abroscopus superciliaris*	R	–	R	R	LC
Mountain Leaftoiler	*Phyllergates cucullatus*	R	–	R	R	LC
Sunda Bush Warbler	*Horornis vulcanius*	–	–	R	R	LC
Manchurian Bush Warbler	*Horornis borealis ?*	–	–	–	V	VU
Bornean Stubtail	*Urosphena whiteheadi*	–	–	R	R	LC
Pnoepygidae						
Pygmy Cupwing	*Pnoepyga pusilla*	R	–	–	–	LC
Acrocephalidae						
Black-browed Reed-warbler	*Acrocephalus bistrigiceps*	M	M	–	–	LC
Oriental Reed-warbler	*Acrocephalus orientalis*	M	M	M	M	LC
Australasian Reed Warbler	*Acrocephalus australis*	–	–	?	?	LC
Thick-billed Warbler	*Acrocephalus aedon*	V	–	–	–	LC

Common English Name	Scientific Name	Peninsular Malaysia	Singapore	Sarawak	Sabah	Global Status
Manchurian Reed Warbler	Acrocephalus tangorum	V	–	–	–	LC
Locustellidae						
Pallas's Grasshopper Warbler	Locustella certhiola	M	M	M	M	LC
Lanceolated Warbler	Locustella lanceolata	M	M	M	M	LC
Middendorf's Grasshopper Warbler	Locustella ochotensis	–	–	M	M	LC
Friendly Grasshopper Warbler	Locustella accentor	–	–	–	R	LC
Striated Grassbird	Megalurus palustris	–	–	R	R	LC
Cisticolidae						
Common Tailorbird	Orthotomus sutorius	R	R	–	–	LC
Dark-necked Tailorbird	Orthotomus atrogularis	R	R	R	R	LC
Ashy Tailorbird	Orthotomus ruficeps	R	R	R	R	LC
Rufous-tailed Tailorbird	Orthotomus sericeus	R	R	R	R	LC
Zitting Cisticola	Cisticola juncidis	R	R	–	–	LC
Hill Prinia	Prinia superciliaris	R	–	–	–	LC
Rufescent Prinia	Prinia rufescens	R	–	–	–	LC
Yellow-bellied Prinia	Prinia flaviventris	R	R	R	R	LC
Sittidae						
Velvet-fronted Nuthatch	Sitta frontalis	R	–	R	R	LC
Blue Nuthatch	Sitta azurea	R	–	–	–	LC
Sturnidae						
Asian Glossy Starling	Aplonis panayensis	R	R	R	R	LC
Golden-crested Myna	Ampeliceps coronatus	X	–	–	–	LC
Common Hill-myna	Gracula religiosa	R	R	R	R	LC
Common Myna	Acridotheres tristis	R	R	F	F	LC
Crested Myna	Acridotheres cristatellus	F	F	–	F	LC
Jungle Myna	Acridotheres fuscus	R	–	–	–	LC
Javan Myna	Acridotheres javanicus	F	F	F	F	LC
White-vented Myna	Acridotheres grandis	F	–	–	–	LC
Black-winged Myna	Acridotheres melanopterus	–	FX	–	–	CR
White-shouldered Starling	Sturnia sinensis	M	M	V	–	LC
Purple-backed Starling	Agropsar sturninus	M	M	V	–	LC
Chestnut-cheeked Starling	Agropsar philippensis	V	V	M	M	LC
Rosy Starling	Pastor roseus	–	V	–	V	LC
Common Starling	Sturnus vulgaris	–	–	–	V	LC
Red-billed Starling	Sturnus sericeus	V	–	–	V	LC
Turdidae						
Everett's Thrush	Zoothera everetti	–	–	R	R	NT
White's Thrush	Zoothera aurea	–	–	–	V	LC
Eurasian Scaly Thrush	Zoothera dauma	V	–	–	–	LC
Chestnut-capped Thrush	Geokichla interpres	R	–	R	R	NT
Orange-headed Thrush	Geokichla citrina	M	M	R	R	LC
Siberian Thrush	Geokichla sibirica	M	M	V	V	LC
Island Thrush	Turdus poliocephalus	–	–	–	R	LC
Japanese Thrush	Turdus cardis	–	–	–	V	LC
Red-throated Thrush	Turdus ruficollis	V	–	–	–	LC
Eyebrowed Thrush	Turdus obscurus	M	M	M	M	LC
Fruithunter	Chlamydochaera jefferyi	–	–	R	R	LC
Muscicapidae						

Common English Name	Scientific Name	Peninsular Malaysia	Singapore	Sarawak	Sabah	Global Status
Oriental Magpie-robin	Copsychus saularis	R	R	R	R	LC
White-rumped Shama	Copsychus malabaricus	R	R	R	R	LC
White-crowned Shama	Copsychus stricklandii	–	–	–	R	
Rufous-tailed Shama	Copsychus pyrropygus	R	–	R	R	NT
Grey-streaked Flycatcher	Muscicapa griseisticta	–	V	V	V	LC
Dark-sided Flycatcher	Muscicapa sibirica	M	M	M	M	LC
Asian Brown Flycatcher	Muscicapa dauurica	M	M	M	M	LC
Brown-streaked Flycatcher	Muscicapa williamsoni	R	M	V	RM	LC
Umber Flycatcher	Muscicapa umbrosa	R	–	R	R	
Ferruginous Flycatcher	Muscicapa ferruginea	M	M	M	M	LC
White-tailed Flycatcher	Cyornis concretus	R	–	R	R	LC
Dayak Jungle-flycatcher	Cyornis montanus	–	–	R	R	LC
Hill Blue Flycatcher	Cyornis caerulatus	R	–	–	–	LC
Tickell's Blue Flycatcher	Cyornis tickelliae	RM	–	–	–	LC
Fulvous-chested Jungle Flycatcher	Cyornis olivaceus	–	–	–	R	LC
Grey-chested Jungle-flycatcher	Cyornis umbratilis	R	–	R	R	NT
Brown-chested Jungle-flycatcher	Cyornis brunneatus	M	M	–	–	VU
Crocker Jungle-flycatcher	Cyornis ruficrissa	–	–	R	R	LC
Large-billed Blue Flycatcher	Cyornis caerulatus	–	–	R	R	VU
Malaysian Blue Flycatcher	Cyornis turcosus	R	–	R	R	NT
Large Blue Flycatcher	Cyornis magnirostris	M	–	–	–	LC
Chinese Blue Flycatcher	Cyornis glaucicomans	M	V	–	–	
Blue-throated Flycatcher	Cyornis rubeculoides	M	V	–	–	LC
Bornean Blue Flycatcher	Cyornis superbus	–	–	R	R	LC
Mangrove Blue Flycatcher	Cyornis rufigastra	R	R	R	R	LC
Ruck's Flycatcher	Cyornis ruckii	?	–	–	–	CR
Pale Blue Flycatcher	Cyornis unicolor	R	–	R	R	LC
Rufous-browed Flycatcher	Anthipes solitaris	R	–	–	–	LC
Rufous-vented Niltava	Niltava sumatrana	R	–	–	–	LC
Large Niltava	Niltava grandis	R	–	–	–	LC
Blue-and-white Flycatcher	Cyanoptila cyanomelana	M	M	M	M	LC
Zappey's Flycatcher	Cyanoptila cumatilis	M	–	–	M	LC
Verditer Warbling-flycatcher	Eumyias thalassinus	R	–	R	R	LC
Indigo Warbling-flycatcher	Eumyias indigo	–	–	R	R	LC
Bornean Shade-dweller	Vauriella gularis	–	–	R	R	LC
Lesser Shortwing	Brachypteryx leucophris	R	–	–	–	LC
Bornean Shortwing	Brachypteryx erythrogyna	–	–	R	R	LC
Siberian Blue Robin	Larvivora cyane	M	M	M	M	LC
Siberian Rubythroat	Calliope calliope	V	–	–	V	LC
Rufous-headed Robin	Luscinia ruficeps	V	–	–	–	EN
White-tailed Robin	Myiomela leucura	R	–	–	–	LC
Siberian Bluetail	Tarsiger cyanurus	–	–	–	V	LC
Chestnut-naped Forktail	Enicurus ruficapillus	R	–	R	R	NT
Slaty-backed Forktail	Enicurus schistaceus	R	–	–	–	LC
Malayan Forktail	Enicurus frontalis	R	–	R	R	LC
Bornean Forktail	Enicurus borneensis	–	–	R	R	
Malaysian Whistling-thrush	Myophonus robinsoni	R	–	–	–	NT
Blue Whistling-thrush	Myophonus caeruleus	RM	–	–	–	LC

Common English Name	Scientific Name	Peninsular Malaysia	Singapore	Sarawak	Sabah	Global Status
Bornean Whistling-thrush	Myophonus borneensis	–	–	R	R	LC
Yellow-rumped Flycatcher	Ficedula zanthopygia	M	M	–	V	LC
Narcissus Flycatcher	Ficedula narcissina	–	–	M	M	LC
Green-backed Flycatcher	Ficedula elisae	M	V	–	–	LC
Mugimaki Flycatcher	Ficedula mugimaki	M	M	M	M	LC
Pygmy Flycatcher	Ficedula hodgsoni	R	–	R	R	LC
Little Pied Flycatcher	Ficedula westermanni	R	–	R	R	LC
Snowy-browed Flycatcher	Ficedula hyperythra	R	–	R	R	LC
Rufous-chested Flycatcher	Ficedula dumetoria	R	–	R	R	NT
Taiga Flycatcher	Ficedula albicilla	M	–	M	M	LC
White-throated Rock-thrush	Monticola gularis	M	V	V	–	LC
Blue Rock-thrush	Monticola solitarius	RM	M	M	M	LC
Stejneger's Stonechat	Saxicola stejnegeri	M	M	V	V	LC
Pied Bushchat	Saxicola caprata	–	–	–	V	LC
Northern Wheatear	Oenanthe oenanthe	–	–	V	–	LC
Chloropseidae						
Lesser Green Leafbird	Chloropsis cyanopogon	R	R	R	R	NT
Greater Green Leafbird	Chloropsis sonnerati	R	R	R	R	VU
Blue-winged Leafbird	Chloropsis moluccensis	R	R	R	–	LC
Bornean Leafbird	Chloropsis kinabaluensis	–	–	–	R	LC
Orange-bellied Leafbird	Chloropsis hardwickii	R	–	–	–	LC
Irenidae						
Asian Fairy-bluebird	Irena puella	R	R	R	R	LC
Dicaeidae						
Yellow-breasted Flowerpecker	Prionochilus maculatus	R	X	R	R	LC
Scarlet-breasted Flowerpecker	Prionochilus thoracicus	R	–	R	R	NT
Crimson-breasted Flowerpecker	Prionochilus percussus	R	–	R	R	LC
Yellow-rumped Flowerpecker	Prionochilus xanthopygius –		R	R	LC	
Brown-backed Flowerpecker	Pachyglossa everetti	R	–	R	R	NT
Modest Flowerpecker	Pachyglossa modesta	R	M	R	R	LC
Orange-bellied Flowerpecker	Dicaeum trigonostigma	R	R	R	R	LC
Plain Flowerpecker	Dicaeum minullum	R	X	R	R	LC
Yellow-vented Flowerpecker	Dicaeum chrysorrheum	R	R	R	R	LC
Fire-breasted Flowerpecker	Dicaeum ignipectus	R	–	–	–	LC
Bornean Flowerpecker	Dicaeum monticolum	–	–	R	R	LC
Spectacled Flowerpecker	Dicaeum sp.	–	–	–	R	
Scarlet-backed Flowerpecker	Dicaeum cruentatum	R	R	R	R	LC
Nectariniidae						
Ruby-cheeked Sunbird	Chalcoparia singalensis	R	–	R	R	LC
Plain Sunbird	Anthreptes simplex	R	V	R	R	LC
Brown-throated Sunbird	Anthreptes malacensis	R	R	R	R	LC
Red-throated Sunbird	Anthreptes rhodolaemus	R	–	R	R	NT
Van Hasselt's Sunbird	Leptocoma brasiliana	R	R	R	R	LC
Copper-throated Sunbird	Leptocoma calcostetha	R	R	R	R	LC
Crimson Sunbird	Aethopyga siparaja	R	R	R	R	LC
Temminck's Sunbird	Aethopyga temminckii	R	–	R	R	LC
Black-throated Sunbird	Aethopyga saturata	R	–	–	–	LC
Ornate Sunbird	Cinnyris ornatus	R	R	R	R	LC

Common English Name	Scientific Name	Peninsular Malaysia	Singapore	Sarawak	Sabah	Global Status
Grey-breasted Spiderhunter	*Arachnothera modesta*	R	X	R	R	LC
Bornean Spiderhunter	*Arachnothera everetti*	–	–	R	R	LC
Streaked Spiderhunter	*Arachnothera magna*	R	–	–	–	LC
Spectacled Spiderhunter	*Arachnothera flavigaster*	R	X	R	R	LC
Yellow-eared Spiderhunter	*Arachnothera chrysogenys*	R	R	R	R	LC
Whitehead's Spiderhunter	*Arachnothera juliae*	–	–	R	R	LC
Little Spiderhunter	*Arachnothera longirostra*	R	R	R	R	LC
Thick-billed Spiderhunter	*Arachnothera crassirostris*	R	R	R	R	LC
Long-billed Spiderhunter	*Arachnothera robusta*	R	–	R	R	LC
Purple-naped Spiderhunter	*Arachnothera hypogrammica*	R	X	R	R	LC
Ploceidae						
Streaked Weaver	*Ploceus manyar*	–	F	–	–	LC
Baya Weaver	*Ploceus philippinus*	R	R	–	–	LC
Estrildidae						
Red Avadavat	*Amandava amandava*	–	F	–	F	LC
Tawny-breasted Parrotfinch	*Erythrura hyperythra*	R	–	R	R	LC
Pin-tailed Parrotfinch	*Erythrura prasina*	R	–	R	R	LC
Scaly-breasted Munia	*Lonchura punctulata*	R	R	R	R	LC
White-rumped Munia	*Lonchura striata*	R	R	–	–	LC
Javan Munia	*Lonchura leucogastroides*	–	F	–	–	LC
White-bellied Munia	*Lonchura leucogastra*	R	–	R	R	LC
Dusky Munia	*Lonchura fuscans*	–	–	R	R	LC
Black-headed Munia	*Lonchura atricapilla*	R	R	R	R	LC
White-headed Munia	*Lonchura maja*	R	R	–	–	LC
Java Sparrow	*Padda oryzivora*	F	F	F	F	VU
Passeridae						
House Sparrow	*Passer domesticus*	–	F	–	–	LC
Plain-backed Sparrow	*Passer flaveolus*	R	–	–	–	LC
Eurasian Tree Sparrow	*Passer montanus*	R	R	R	R	LC
Motacillidae						
Forest Wagtail	*Dendronanthus indicus*	M	M	M	M	LC
Eastern Yellow Wagtail	*Motacilla tschutschensis*	M	M	M	M	LC
Citrine Wagtail	*Motacilla citreola*	V	V	–	–	LC
Grey Wagtail	*Motacilla cinerea*	M	M	M	M	LC
White Wagtail	*Motacilla alba*	M	M	M	M	LC
Olive-backed Pipit	*Anthus hodgsoni*	M	–	M	M	LC
Blyth's Pipit	*Anthus godlewskii*	V	–	–	–	LC
Red-throated Pipit	*Anthus cervinus*	M	M	M	M	LC
Pechora Pipit	*Anthus gustavi*	–	–	M	M	LC
Paddyfield Pipit	*Anthus rufulus*	R	R	R	R	LC
Richard's Pipit	*Anthus richardi*	V?	–	M	M	LC
Fringillidae						
Malay Bullfinch	*Pyrrhula waterstradti*	R	–	–	–	VU
Emberizidae						
Chestnut-eared Bunting	*Emberiza fucata*	V	–	–	–	LC
Little Bunting	*Emberiza pusilla*	–	–	V	V	LC
Chestnut Bunting	*Emberiza rutila*	V	–	–	–	LC
Yellow-breasted Bunting	*Emberiza aureola*	M	M	V	V	EN

FURTHER READING

Bowden, D. (2005). *Globetrotter Visitor's Guide to Taman Negara, Malaysia's Premier National Park.* New Holland Publishers (UK) Ltd.

Bransbury, J. (1993). *A Birdwatcher's Guide to Malaysia.* Waymark Publishing.

Davison, Payne & Gumal. (2013). *Wild Malaysia.* John Beaufoy Publishing

Eaton, J.A., van Balen, B., Brickle, N.W. and Rheindt, F.E. 2016. *Birds of the Indonesian Archipelago. Greater Sundas and Wallacea.* Lynx Edicions, Barcelona.

Francis, C.M. (2008). *A Field Guide to the Mammals of South-east Asia.* New Holland Publishers (UK) Ltd.

Jeyarajasingam, A. and Pearson, A. (1999). *A Field Guide to Birds of Peninsular Malaysia and Singapore.* Oxford University Press.

Lim, K.S. (2009). *The Avifauna of Singapore.* Nature Society (Singapore).

MacKinnon, J. and Phillipps, K. (1993). *A Field Guide to the Birds of Borneo, Sumatra, Java and Bali.* Oxford University Press Inc.

Mann, C.F. (2008). *The Birds of Borneo: an Annotated Checklist.* British Ornithologists' Union.

Myers, S. (2009). *A Field Guide to the Birds of Borneo.* Talisman and New Holland Publishers (UK) Ltd.

Payne, J. (2010). *Wild Sabah: the Magnificent Wildlife and Rainforests of Malaysian Borneo.* John Beaufoy Publishing.

Phillipps, A. and Liew, F. (2000). *Globetrotter Visitor's Guide to Kinabalu Park, Sabah, Malaysian Borneo.* New Holland Publishers (UK) Ltd.

Phillipps, Q. and Phillipps, K. (2009). *Phillipps' Field Guide to the Birds of Borneo.* Beaufoy Books.

Robson, C. (2008). *A Field Guide to the Birds of South-east Asia.* New Holland Publishers (UK) Ltd.

Smythies, B.E. (1999). *The Birds of Borneo.* 4th edn. Natural History Publications (Borneo) Sdn. Bhd. and the Sabah Society.

Strange, M. (2004). *Birds of Fraser's Hill.* Nature's Niche Pte Ltd.

Strange, M. and Jeyarajasingam, A. (1999). *A Photographic Guide to the Birds of Peninsular Malaysia and Singapore.* Sun Tree Publishing.

Strange, M. and Yong, D. (2006). *Birds of Taman Negara.* Draco Publishing and Distribution Pte Ltd.

Tan, H.T.W., Chou, L.M., Yeo, D.C.J. and Ng, P.K.L. (2007). *The Natural Heritage of Singapore.* Prentice Hall.

Wang, L.K. and Hails, C.J. (2007). An Annotated Checklist of the Birds of Singapore. *Raffles Bulletin of Zoology,* Supplement 15.

Wells, D.R. (1999). *The Birds of the Thai-Malay Peninsula. Volume One: Non-Passerines.* Academic Press.

Wells, D.R. (2007). *The Birds of the Thai-Malay Peninsula. Volume Two: Passerines.* Christopher Helm.

Whitmore, T.C. (1984). *Tropical Rain Forests of the Far East.* Oxford University Press.

Wong, T.S. (2014). *A Naturalist's Guide to the Birds of Borneo.* John Beaufoy Publishing.

Yong, Lim & Lee. (2014). *A Naturalist's Guide to the Birds of Singapore.* John Beaufoy Publishing.

USEFUL CONTACTS

Borneo Bird Club
http//:borneobirdclub.blogspot.com

Malaysian Nature Society
JKR 641, Jalan Kelantan
Bukit Persekutuan
50480 Kuala Lumpur
Malaysia
www.mns.org.my

Nature Society (Singapore)
510 Geylang Road #02–05
The Sunflower
Singapore 389466
www.nss.org.sg

Oriental Bird Club
PO Box 324
Bedford MK42 0WG
UK
www.orientalbirdclub.org
www.orientalbirdimages.org

■ INDEX ■